Solutions Manual
to Accompany

Accelerated Studies in Physics and Chemistry

second edition

Rebekah L. Mays and John D. Mays

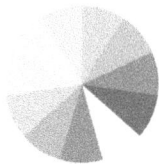

CENTRIPETAL PRESS
CLASSICAL ACADEMIC PRESS

Camp Hill, Pennsylvania
2021

CENTRIPETAL PRESS
CLASSICAL ACADEMIC PRESS

Solutions Manual to Accompany Accelerated Studies in Physics and Chemistry

© Classical Academic Press®, 2021

Edition 2.0

ISBN: 978-1-7326384-1-9

Classical Academic Press

515 S. 32nd Street

Camp Hill, PA 17011

www.ClassicalAcademicPress.com/Novare/

Cover design: John D. Mays

Contents

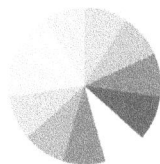

CENTRIPETAL PRESS
CLASSICAL ACADEMIC PRESS

Acknowledgement

I (John) wish to express my gratitude to my daughter Rebekah Mays for carefully and meticulously compiling these solutions. It may have been a walk down memory lane for her: she took this course back in 04–05 when she was in 9th grade. (I was the teacher.)

Any errors that remain in this volume are my own responsibility.

Preface

This solutions manual contains detailed solutions for all the computational problems contained in my text *Accelerated Studies in Physics and Chemistry*, second edition. Teachers and students using that text should find this manual to be a valuable resource.

All results are shown using the correct number of significant digits for the data given in the problem. (In some cases, an extra digit might be included in data tables use for constructing graphs in order to improve the appearance of the graph.) Guidelines for determining the correct number of significant digits for a given problem are given in Section 2.1.5 in the text.

When comparing your results to the results shown here and to those in the text, keep in mind that the last digit is always uncertain because of the way significant digits in measurements are defined. When two results match except for a small difference in the most precise digit, we say that the results match. Because of rounding in calculators, it is not uncommon for results shown here to differ from the answer key in the text or from your result by one or two in the most precise digit.

We have checked and double checked the solutions to make them as accurate as possible. However, in any manual of this kind it is likely that errors remain. If you find an error, we would be much obliged if you would inform us of it by sending an email to info@centripetalpress.com.

Chapter 2

Unit Conversions

1.

$$1750 \text{ m} \cdot \frac{100 \text{ cm}}{1 \text{ m}} \cdot \frac{1 \text{ in}}{2.54 \text{ cm}} \cdot \frac{1 \text{ ft}}{12 \text{ in}} = 5740 \text{ ft}$$

2.

$$3.54 \text{ g} \cdot \frac{1 \text{ kg}}{1000 \text{ g}} = 0.00354 \text{ kg}$$

3.

$$41.11 \text{ mL} \cdot \frac{1 \text{ L}}{1000 \text{ mL}} = 0.04111 \text{ L}$$

4.

$$7 \times 10^8 \text{ m} \cdot \frac{100 \text{ cm}}{1 \text{ m}} \cdot \frac{1 \text{ in}}{2.54 \text{ cm}} \cdot \frac{1 \text{ ft}}{12 \text{ in}} \cdot \frac{1 \text{ mi}}{5280 \text{ ft}} = 4 \times 10^5 \text{ mi}$$

5.

$$1.5499 \times 10^{-12} \text{ mm} \cdot \frac{1 \text{ m}}{1000 \text{ mm}} = 1.5499 \times 10^{-15} \text{ m}$$

6.

$$750 \text{ cm}^3 \cdot \frac{1 \text{ mL}}{1 \text{ cm}^3} \cdot \frac{1 \text{ L}}{1000 \text{ mL}} \cdot \frac{1 \text{ m}^3}{1000 \text{ L}} = 7.5 \times 10^{-4} \text{ m}^3$$

7.

$$2.9979 \times 10^8 \frac{\text{m}}{\text{s}} \cdot \frac{100 \text{ cm}}{1 \text{ m}} \cdot \frac{1 \text{ in}}{2.54 \text{ cm}} \cdot \frac{1 \text{ ft}}{12 \text{ in}} = 9.8356 \times 10^8 \frac{\text{ft}}{\text{s}}$$

8.

$$168 \text{ hr} \cdot \frac{60 \text{ min}}{1 \text{ hr}} \cdot \frac{60 \text{ s}}{1 \text{ min}} = 605{,}000 \text{ s}$$

9.

$$5570 \frac{\text{kg}}{\text{m}^3} \cdot \frac{1000 \text{ g}}{1 \text{ kg}} \cdot \frac{1 \text{ m}^3}{1000 \text{ L}} \cdot \frac{1 \text{ L}}{1000 \text{ mL}} \cdot \frac{1 \text{ mL}}{1 \text{ cm}^3} = 5.57 \frac{\text{g}}{\text{cm}^3}$$

10.

$$45\,\frac{\text{gal}}{\text{s}}\cdot\frac{3.786\text{ L}}{1\text{ gal}}\cdot\frac{1\text{ m}^3}{1000\text{ L}}\cdot\frac{60\text{ s}}{1\text{ min}}=1.0\times10^1\,\frac{\text{m}^3}{\text{min}}$$

11.

$$600,000\,\frac{\text{ft}^3}{\text{s}}\cdot\frac{(0.3048\text{ m})^3}{1\text{ ft}^3}\cdot\frac{1000\text{ L}}{1\text{ m}^3}\cdot\frac{60\text{ s}}{1\text{ min}}\cdot\frac{60\text{ min}}{1\text{ hr}}=6\times10^{10}\,\frac{\text{L}}{\text{hr}}$$

12.

$$5200\text{ mL}\cdot\frac{1\text{ L}}{1000\text{ mL}}\cdot\frac{1\text{ m}^3}{1000\text{ L}}=5.2\times10^{-3}\text{ m}^3$$

13.

$$5.65\times10^2\text{ mm}^2\cdot\frac{1\text{ cm}}{10\text{ mm}}\cdot\frac{1\text{ cm}}{10\text{ mm}}\cdot\frac{1\text{ in}}{2.54\text{ cm}}\cdot\frac{1\text{ in}}{2.54\text{ cm}}=0.876\text{ in}^2$$

14.

$$32.16\,\frac{\text{ft}}{\text{s}^2}\cdot\frac{12\text{ in}}{1\text{ ft}}\cdot\frac{2.54\text{ cm}}{1\text{ in}}\cdot\frac{1\text{ m}}{100\text{ cm}}=9.802\,\frac{\text{m}}{\text{s}^2}$$

15.

$$5.001\,\frac{\mu\text{g}}{\text{s}}\cdot\frac{1\text{ g}}{10^6\,\mu\text{g}}\cdot\frac{1\text{ kg}}{1000\text{ g}}\cdot\frac{60\text{ s}}{1\text{ min}}=3.001\times10^{-4}\,\frac{\text{kg}}{\text{min}}$$

16.

$$4.771\,\frac{\text{g}}{\text{mL}}\cdot\frac{1\text{ kg}}{1000\text{ g}}\cdot\frac{1000\text{ mL}}{1\text{ L}}\cdot\frac{1000\text{ L}}{1\text{ m}^3}=4771\,\frac{\text{kg}}{\text{m}^3}$$

17.

$$13.6\,\frac{\text{g}}{\text{cm}^3}\cdot\frac{1000\text{ mg}}{1\text{ g}}\cdot\frac{100\text{ cm}}{1\text{ m}}\cdot\frac{100\text{ cm}}{1\text{ m}}\cdot\frac{100\text{ cm}}{1\text{ m}}=1.36\times10^{10}\,\frac{\text{mg}}{\text{m}^3}$$

18.

$$93,000,000\text{ mi}\cdot\frac{5280\text{ ft}}{1\text{ mi}}\cdot\frac{0.3048\text{ m}}{1\text{ ft}}\cdot\frac{100\text{ cm}}{1\text{ m}}=1.5\times10^{13}\text{ cm}$$

19.

$$65\,\frac{\text{mi}}{\text{hr}}\cdot\frac{5280\text{ ft}}{1\text{ mi}}\cdot\frac{0.3048\text{ m}}{1\text{ ft}}\cdot\frac{1\text{ hr}}{60\text{ min}}\cdot\frac{1\text{ min}}{60\text{ s}}=29\,\frac{\text{m}}{\text{s}}$$

20.

$$633 \text{ nm} \cdot \frac{1 \text{ m}}{1 \times 10^9 \text{ nm}} \cdot \frac{100 \text{ cm}}{1 \text{ m}} \cdot \frac{1 \text{ in}}{2.54 \text{ cm}} = 2.49 \times 10^{-5} \text{ in}$$

21.

$$0.05015 \cdot 3.00 \times 10^8 \ \frac{\text{m}}{\text{s}} \cdot \frac{60 \text{ s}}{1 \text{ min}} \cdot \frac{60 \text{ min}}{1 \text{ hr}} \cdot \frac{1 \text{ ft}}{0.3048 \text{ m}} \cdot \frac{1 \text{ mi}}{5280 \text{ ft}} = 3.37 \times 10^7 \ \frac{\text{mi}}{\text{hr}}$$

22.

$T_F = 98.6°\text{F}$

$T_C = ?$

$$T_C = \frac{5}{9}(T_F - 32) = \frac{5}{9}(98.6°\text{F} - 32) = 37.0°\text{C}$$

23.

$T_C = 50.0°\text{C}$

$T_F = ?$

$$T_C = \frac{5}{9}(T_F - 32)$$

$$T_F = \frac{9}{5}T_C + 32 = \frac{9}{5}(50.0°\text{C}) + 32 = 122°\text{F}$$

24.

$$t = 1 \text{ yr} \cdot \frac{365 \text{ days}}{1 \text{ year}} \cdot \frac{24 \text{ hr}}{1 \text{ day}} \cdot \frac{60 \text{ min}}{1 \text{ hr}} \cdot \frac{60 \text{ s}}{1 \text{ min}} = 31,540,000 \text{ s}$$

$$v = c = 3.00 \times 10^8 \ \frac{\text{m}}{\text{s}}$$

$d = ?$

$$v = \frac{d}{t}$$

$$d = vt$$

$$d = 3.00 \times 10^8 \ \frac{\text{m}}{\text{s}} \cdot 31,540,000 \text{ s} = 9.46 \times 10^{15} \text{ m (this is one lt-yr expressed in m.)}$$

$$4.3 \text{ lt-yr} = 4.3 \cdot 9.46 \times 10^{15} \text{ m} = 4.07 \times 10^{16} \text{ m} \cdot \frac{1 \text{ km}}{1000 \text{ m}} = 4.1 \times 10^{13} \text{ km}$$

Motion Study Questions Set 1

1.

$$d = 25.1 \text{ mi} \cdot \frac{5280 \text{ ft}}{1 \text{ mi}} \cdot \frac{0.3048 \text{ m}}{1 \text{ ft}} = 4.04 \times 10^4 \text{ m}$$

$$t = 0.50 \text{ hr} \cdot \frac{60 \text{ min}}{1 \text{ hr}} \cdot \frac{60 \text{ s}}{1 \text{ min}} = 1800 \text{ s}$$

$$v = ?$$

$$v = \frac{d}{t} = \frac{4.04 \times 10^4 \text{ m}}{1800 \text{ s}} = 22 \frac{\text{m}}{\text{s}}$$

2.

$$22 \frac{\text{m}}{\text{s}} \cdot \frac{1 \text{ km}}{1000 \text{ m}} \cdot \frac{60 \text{ s}}{1 \text{ min}} \cdot \frac{60 \text{ min}}{1 \text{ hr}} = 79 \frac{\text{km}}{\text{hr}}$$

3.

$$t = 4.25 \text{ hr} \cdot \frac{3600 \text{ s}}{\text{hr}} = 15,300 \text{ s}$$

$$v = 5.0000 \frac{\text{km}}{\text{hr}} \cdot \frac{1000 \text{ m}}{\text{km}} \cdot \frac{1 \text{ hr}}{3600 \text{ s}} = 1.389 \frac{\text{m}}{\text{s}}$$

$$d = ?$$

$$v = \frac{d}{t}$$

$$d = vt$$

$$d = 1.389 \frac{\text{m}}{\text{s}} \cdot 15,300 \text{ s} = 21,300 \text{ m} \cdot \frac{1 \text{ km}}{1000 \text{ m}} = 21.3 \text{ km}$$

4.

$$21.3 \text{ km} \cdot \frac{1000 \text{ m}}{1 \text{ km}} \cdot \frac{1 \text{ ft}}{0.3048 \text{ m}} \cdot \frac{1 \text{ mi}}{5,280 \text{ ft}} = 13.2 \text{ mi}$$

5.

$$150.0 \frac{\text{mi}}{\text{hr}} \cdot \frac{5280 \text{ ft}}{1 \text{ mi}} \cdot \frac{0.3048 \text{ m}}{1 \text{ ft}} \cdot \frac{1 \text{ km}}{1000 \text{ m}} = 241.4 \frac{\text{km}}{\text{hr}}$$

6.

$$v = 150.0 \,\frac{\text{mi}}{\text{hr}} \cdot \frac{1609 \text{ m}}{\text{mi}} \cdot \frac{1 \text{ hr}}{3600 \text{ s}} = 67.04 \,\frac{\text{m}}{\text{s}}$$

$$d = 10.0 \text{ mi} \cdot \frac{1609 \text{ m}}{\text{mi}} = 16{,}090 \text{ m}$$

$$t = ?$$

$$v = \frac{d}{t}$$

$$t = \frac{d}{v} = \frac{16{,}090 \text{ m}}{67.04 \,\dfrac{\text{m}}{\text{s}}} = 240.0 \text{ s} \cdot \frac{1 \text{ min}}{60 \text{ s}} = 4.00 \text{ min}$$

7.

$$d = 3.0 \text{ km} \cdot \frac{1000 \text{ m}}{1 \text{ km}} = 3.0 \times 10^3 \text{ m}$$

$$t = 1 \text{ hr } 20.0 \text{ min} = 80.0 \text{ min} \cdot \frac{60 \text{ s}}{1 \text{ min}} = 4.80 \times 10^3 \text{ s}$$

$$v = ?$$

$$v = \frac{d}{t} = \frac{3.0 \times 10^3 \text{ m}}{4.80 \times 10^3 \text{ s}} = 0.63 \,\frac{\text{m}}{\text{s}}$$

8.

$$v_i = 0$$

$$v_f = 45 \,\frac{\text{mi}}{\text{hr}} \cdot \frac{1 \text{ hr}}{60 \text{ min}} \cdot \frac{1 \text{ min}}{60 \text{ s}} \cdot \frac{5{,}280 \text{ ft}}{1 \text{ mi}} \cdot \frac{0.3048 \text{ m}}{1 \text{ ft}} = 20.1 \,\frac{\text{m}}{\text{s}}$$

$$t = 36 \text{ s}$$

$$a = ?$$

$$a = \frac{v_f - v_i}{t} = \frac{20.1 \,\dfrac{\text{m}}{\text{s}} - 0}{36 \text{ s}} = 0.56 \,\frac{\text{m}}{\text{s}^2}$$

9.

$$v_i = 31 \; \frac{m}{s}$$

$$t = 17 \; s$$

$$v_f = 22 \; \frac{m}{s}$$

$$a = ?$$

$$a = \frac{v_f - v_i}{t} = \frac{22 \; \frac{m}{s} - 31 \; \frac{m}{s}}{17 \; s} = -0.53 \; \frac{m}{s^2}$$

10.

$$d = 14.5 \; m$$

$$v = c = 3.00 \times 10^8 \; \frac{m}{s}$$

$$t = ?$$

$$v = \frac{d}{t}$$

$$t = \frac{d}{v} = \frac{14.5 \; m}{3.00 \times 10^8 \; \frac{m}{s}} = 4.83 \times 10^{-8} \; s \cdot \frac{1 \times 10^9 \; ns}{s} = 48.3 \; ns$$

11.

$$v_i = 0$$

$$v_f = 0.80 \cdot 3.00 \times 10^8 \; \frac{m}{s} = 2.40 \times 10^8 \; \frac{m}{s}$$

$$t = 18 \; hr \; 6 \; min \; 45 \; s = 64,800 \; s + 360 \; s + 45 \; s = 65,205 \; s$$

$$a = ?$$

$$a = \frac{v_f - v_i}{t} = \frac{2.40 \times 10^8 \; \frac{m}{s} - 0}{65,205 \; s} = 3680 \; \frac{m}{s^2}$$

12.

$$d = 8.96 \times 10^9 \text{ km} \cdot \frac{1000 \text{ m}}{1 \text{ km}} = 8.96 \times 10^{12} \text{ m}$$

$$v = 3.45 \times 10^5 \ \frac{\text{m}}{\text{s}}$$

$$t = ?$$

$$v = \frac{d}{t}$$

$$t = \frac{d}{v} = \frac{8.96 \times 10^{12} \text{ m}}{3.45 \times 10^5 \ \frac{\text{m}}{\text{s}}} = 2.597 \times 10^7 \text{ s} \cdot \frac{1 \text{ hr}}{3600 \text{ s}} \cdot \frac{1 \text{ day}}{24 \text{ hr}} = 301 \text{ days}$$

13.

$$a = 5.556 \times 10^6 \ \frac{\text{cm}}{\text{s}^2} \cdot \frac{1 \text{ m}}{100 \text{ cm}} = 5.556 \times 10^4 \ \frac{\text{m}}{\text{s}^2}$$

$$t = 45 \text{ ms} \cdot \frac{1 \text{ s}}{1000 \text{ ms}} = 4.5 \times 10^{-2} \text{ s}$$

$$v_i = 0$$

$$v_f = ?$$

$$a = \frac{v_f - v_i}{t}$$

$$v_f = at + v_i = (5.556 \times 10^4 \ \frac{\text{m}}{\text{s}^2})(4.5 \times 10^{-2} \text{ s}) + (0 \ \frac{\text{m}}{\text{s}}) = 2.5 \times 10^3 \ \frac{\text{m}}{\text{s}}$$

14.

$$v_i = 4.005 \times 10^3 \ \frac{\text{m}}{\text{s}}$$

$$a = 23.1 \ \frac{\text{m}}{\text{s}^2}$$

$$t = 13.5 \text{ s}$$

$$v_f = ?$$

$$a = \frac{v_f - v_i}{t}$$

$$v_f = at + v_i = (23.1 \ \frac{\text{m}}{\text{s}^2} \cdot 13.5 \text{ s}) + 4.005 \times 10^3 \ \frac{\text{m}}{\text{s}} = 4.32 \times 10^3 \ \frac{\text{m}}{\text{s}}$$

15.

$$v = c = 2.9979 \times 10^8 \ \frac{m}{s}$$

$$d = 1.4965 \times 10^8 \ km \cdot \frac{1000 \ m}{1 \ km} = 1.4965 \times 10^{11} \ m$$

$$t = ?$$

$$v = \frac{d}{t}$$

$$t = \frac{d}{v} = \frac{1.4965 \times 10^{11} \ m}{2.9979 \times 10^8 \ \frac{m}{s}} = 499.18 \ s \cdot \frac{1 \ min}{60 \ s} = 8.3197 \ min$$

Chapter 3

Newton's Second Law Practice Problems

1.

$m = 1880 \text{ kg}$

$a = 1.50 \ \dfrac{\text{m}}{\text{s}^2}$

$F = ?$

$a = \dfrac{F}{m}$

$F = ma = 1880 \text{ kg} \cdot 1.50 \ \dfrac{\text{m}}{\text{s}^2} = 2820 \text{ N}$

2.

$m = 188.4 \text{ g} \cdot \dfrac{1 \text{ kg}}{1000 \text{ g}} = 0.1884 \text{ kg}$

$g = 9.80 \ \dfrac{\text{m}}{\text{s}^2}$

$F_w = ?$

$F_w = 0.1884 \text{ kg} \cdot 9.80 \ \dfrac{\text{m}}{\text{s}^2} = 1.85 \text{ N}$

3.

$F = 250.0 \text{ N}$

$m = 144{,}000 \text{ mg} \cdot \dfrac{1 \text{ g}}{1000 \text{ mg}} \cdot \dfrac{1 \text{ kg}}{1000 \text{ g}} = 0.144 \text{ kg}$

$a = ?$

$a = \dfrac{F}{m} = \dfrac{250.0 \text{ N}}{0.144 \text{ kg}} = 1740 \ \dfrac{\text{m}}{\text{s}^2}$

4.

$$a = 2.3 \ \frac{m}{s^2}$$

$$F = 230,000 \ N$$

$$m = ?$$

$$a = \frac{F}{m}$$

$$m = \frac{F}{a} = \frac{230,000 \ N}{2.3 \ \frac{m}{s^2}} = 1.0 \times 10^5 \ kg$$

5.

$$a = 0.0022 \ \frac{mi}{hr^2} \cdot \frac{5280 \ ft}{1 \ mi} \cdot \frac{0.3048 \ m}{1 \ ft} \cdot \frac{1 \ hr}{3600 \ s} \cdot \frac{1 \ hr}{3600 \ s} = 2.732 \times 10^{-7} \ \frac{m}{s^2}$$

$$m = 2.2 \ Mg \cdot \frac{1 \times 10^6 \ g}{1 \ Mg} \cdot \frac{1 \ kg}{1000 \ g} = 2.2 \times 10^3 \ kg$$

$$F = ?$$

$$a = \frac{F}{m}$$

$$F = ma = 2.2 \times 10^3 \ kg \cdot 2.732 \times 10^{-7} \ \frac{m}{s^2} = 6.0 \times 10^{-4} \ N$$

6.

$$F_w = 125.1 \ lb \cdot \frac{4.45 \ N}{1 \ lb} = 556.7 \ N$$

$$g = 9.80 \ \frac{m}{s^2}$$

$$m = ?$$

$$F_w = mg$$

$$m = \frac{F_w}{g} = \frac{556.7 \ N}{9.80 \ \frac{m}{s^2}} = 56.8 \ kg$$

7.

$m = 56.8$ kg

$$F_w = 20.9 \text{ lb} \cdot \frac{4.45 \text{ N}}{1 \text{ lb}} = 93.01 \text{ N}$$

$g_m = ?$

$F_w = mg_m$

$$g_m = \frac{F_w}{m} = \frac{93.01 \text{ N}}{56.8 \text{ kg}} = 1.64 \frac{\text{m}}{\text{s}^2}$$

8.

$v_i = 0$

$$v_f = 125.0 \frac{\text{m}}{\text{s}}$$

$$t = 22.00 \text{ ms} \cdot \frac{1 \text{ s}}{1000 \text{ ms}} = 2.200 \times 10^{-2} \text{ s}$$

$F = 142.0$ N

$m = ?$

$$a = \frac{v_f - v_i}{t} = \frac{125.0 \frac{\text{m}}{\text{s}} - 0}{2.200 \times 10^{-2} \text{ s}} = 5681.8 \frac{\text{m}}{\text{s}^2}$$

$$a = \frac{F}{m}$$

$$m = \frac{F}{a} = \frac{142.0 \text{ N}}{5681.8 \frac{\text{m}}{\text{s}^2}} = 0.024992 \text{ kg} \cdot \frac{1000 \text{ g}}{1 \text{ kg}} = 24.99 \text{ g}$$

9.

$m = 4.5 \text{ kg}$

$v_i = 0$

$$v_f = 8.00 \ \frac{\text{mi}}{\text{hr}} \cdot \frac{1 \text{ hr}}{3600 \text{ s}} \cdot \frac{5280 \text{ ft}}{1 \text{ mi}} \cdot \frac{0.3048 \text{ m}}{1 \text{ ft}} = 3.6 \ \frac{\text{m}}{\text{s}}$$

$$t = 500 \text{ ms} \cdot \frac{1 \text{ s}}{1000 \text{ ms}} = 0.5 \text{ s}$$

$F = ?$

$$a = \frac{v_f - v_i}{t} = \frac{3.6 \ \frac{\text{m}}{\text{s}} - 0}{0.5 \text{ s}} = 7.2 \ \frac{\text{m}}{\text{s}^2}$$

$$a = \frac{F}{m}$$

$$F = ma = 4.5 \text{ kg} \cdot 7.2 \ \frac{\text{m}}{\text{s}^2} = 30 \text{ N}$$

10.

$$v_i = 2500.0 \ \frac{\text{km}}{\text{hr}} \cdot \frac{1 \text{ hr}}{3600 \text{ s}} \cdot \frac{1000 \text{ m}}{1 \text{ km}} = 694.4 \ \frac{\text{m}}{\text{s}}$$

$t = 8.000 \text{ s}$

$F = 45,450 \text{ N}$

$$v_f = 2750 \ \frac{\text{km}}{\text{hr}} \cdot \frac{1000 \text{ m}}{1 \text{ km}} \cdot \frac{1 \text{ hr}}{3600 \text{ s}} = 763.9 \ \frac{\text{m}}{\text{s}}$$

$m = ?$

$$a = \frac{v_f - v_i}{t} = \frac{763.9 \ \frac{\text{m}}{\text{s}} - 694.4 \ \frac{\text{m}}{\text{s}}}{8.000 \text{ s}} = 8.688 \ \frac{\text{m}}{\text{s}^2}$$

$$a = \frac{F}{m}$$

$$m = \frac{F}{a} = \frac{45,450 \text{ N}}{8.688 \ \frac{\text{m}}{\text{s}^2}} = 5230 \text{ kg}$$

11.

$$m = 166 \text{ g} \cdot \frac{1 \text{ kg}}{1000 \text{ g}} = 0.166 \text{ kg}$$

$F = 0.0450 \text{ N}$

$v_i = 0$

$t = 2.1 \text{ s}$

$v_f = ?$

$$a = \frac{F}{m} = \frac{0.0450 \text{ N}}{0.166 \text{ kg}} = 0.2711 \frac{\text{m}}{\text{s}^2}$$

$$a = \frac{v_f - v_i}{t}$$

$$v_f = at + v_i = \left(0.2711 \frac{\text{m}}{\text{s}^2} \cdot 2.1 \text{ s}\right) + 0 = 0.57 \frac{\text{m}}{\text{s}}$$

12.

$$m = 1.673 \times 10^{-18} \text{ μg} \cdot \frac{1 \text{ g}}{1 \times 10^6 \text{ μg}} \cdot \frac{1 \text{ kg}}{1000 \text{ g}} = 1.673 \times 10^{-27} \text{ kg}$$

$v_i = 0$

$$v_f = c \cdot 0.0005 = 3.00 \times 10^8 \cdot 0.0005 = 1.50 \times 10^5 \frac{\text{m}}{\text{s}}$$

$$t = 455 \text{ ns} \cdot \frac{1 \text{ s}}{1 \times 10^9 \text{ ns}} = 4.55 \times 10^{-7} \text{ s}$$

$F = ?$

$$a = \frac{v_f - v_i}{t} = \frac{1.50 \times 10^5 \frac{\text{m}}{\text{s}} - 0}{4.55 \times 10^{-7} \text{ s}} = 3.30 \times 10^{11} \frac{\text{m}}{\text{s}^2}$$

$$a = \frac{F}{m}$$

$$F = ma = 1.673 \times 10^{-27} \text{ kg} \cdot 3.30 \times 10^{11} \frac{\text{m}}{\text{s}^2} = 5.52 \times 10^{-16} \text{ N} \cdot \frac{1 \text{ GN}}{1 \times 10^9 \text{ N}} = 5.52 \times 10^{-25} \text{ GN}$$

13.

$$m = 6.548 \text{ Gg} \cdot \frac{1 \times 10^9 \text{ g}}{1 \text{ Gg}} \cdot \frac{1 \text{ kg}}{1000 \text{ g}} = 6.548 \times 10^6 \text{ kg}$$

$$v_i = 8.35 \frac{\text{mi}}{\text{hr}} \cdot \frac{1609 \text{ m}}{1 \text{ mi}} \cdot \frac{1 \text{ hr}}{3600 \text{ s}} = 3.732 \frac{\text{m}}{\text{s}}$$

$$v_f = 0$$

$$t = 0.288 \text{ min} \cdot \frac{60 \text{ s}}{1 \text{ min}} = 17.28 \text{ s}$$

$$F = ?$$

$$a = \frac{v_f - v_i}{t} = \frac{0 - 3.732 \frac{\text{m}}{\text{s}}}{17.28 \text{ s}} = -0.216 \frac{\text{m}}{\text{s}^2}$$

$$a = \frac{F}{m}$$

$$F = ma$$

$$F = 6.548 \times 10^6 \text{ kg} \cdot \left(-0.216 \frac{\text{m}}{\text{s}^2} \right) = -1.41 \times 10^6 \text{ N (The negative sign indicates}$$

that the force is a stopping force, i.e., it opposes the motion.)

14.

$$v_i = 3.5 \frac{\text{cm}}{\text{s}} \cdot \frac{1 \text{ m}}{100 \text{ cm}} = 0.035 \frac{\text{m}}{\text{s}}$$

$$v_f = 18.5 \frac{\text{cm}}{\text{s}} \cdot \frac{1 \text{ m}}{100 \text{ cm}} = 0.185 \frac{\text{m}}{\text{s}}$$

$$t = 220 \text{ ms} \cdot \frac{1 \text{ s}}{1000 \text{ ms}} = 0.22 \text{ s}$$

$$a = ?$$

$$a = \frac{v_f - v_i}{t} = \frac{0.185 \frac{\text{m}}{\text{s}} - 0.035 \frac{\text{m}}{\text{s}}}{0.22 \text{ s}} = 0.68 \frac{\text{m}}{\text{s}^2}$$

15.a.

$m = 45{,}500$ kg

$v_i = 0 \; \dfrac{\text{m}}{\text{s}}$

$v_f = 55 \; \dfrac{\text{m}}{\text{s}}$

$t = 6.4$ s

$a = ?$

$F = ?$

$$a = \frac{v_f - v_i}{t} = \frac{55 \; \frac{\text{m}}{\text{s}} - 0}{6.4 \; \text{s}} = 8.59 \; \frac{\text{m}}{\text{s}^2}$$

$$\boxed{a = 8.6 \; \frac{\text{m}}{\text{s}^2}}$$

15.b.

$a = \dfrac{F}{m}$

$$F = ma = 45{,}500 \; \text{kg} \cdot 8.59 \; \frac{\text{m}}{\text{s}^2} = 3.9 \times 10^5 \; \text{N}$$

16.

$$m = 8.5 \; \text{g} \cdot \frac{1 \; \text{kg}}{1000 \; \text{g}} = 0.0085 \; \text{kg}$$

$a = 18{,}500 \; \dfrac{\text{m}}{\text{s}^2}$

$F = ?$

$a = \dfrac{F}{m}$

$$F = ma = 0.0085 \; \text{kg} \cdot 18{,}500 \; \frac{\text{m}}{\text{s}^2} = 160 \; \text{N}$$

Chapter 4

Note: The values chosen for each activity are examples only, and so answers will vary between students' papers. Students can choose any values for their calculations, so long as their chosen values are within the parameters of each activity.

Activity 1. How does the area of a triangle vary with its height, if all else is held constant?

1. What is the relation for the area of a triangle?

$$A = \frac{bh}{2}$$

2. What are the two key variables that need to be compared in this activity?

A and h

3. Which variable is the independent variable and which is the dependent variable?

A is dependent; h is independent.

4. After combining and normalizing all non-essential variables and constants, what is the expression relating area to height?

$A \propto h$

5. Select any value other than 2 to use for the base of a triangle. Using this value for the base, choose several values for the height of the triangle and calculate the area of the triangle for each height. Enter all these in a table of values.

$$A = \frac{bh}{2}$$

$b = 4.0$ m

$h = 2.0$ m, 4.0 m, 6.0 m, 8.0 m, 10.0 m

$$A = \frac{4.0 \text{ m} \cdot 2.0 \text{ m}}{2} = 4.0 \text{ m}^2$$

$$A = \frac{4.0 \text{ m} \cdot 4.0 \text{ m}}{2} = 8.0 \text{ m}^2$$

$$A = \frac{4.0 \text{ m} \cdot 6.0 \text{ m}}{2} = 12.0 \text{ m}^2$$

$$A = \frac{4.0 \text{ m} \cdot 8.0 \text{ m}}{2} = 16.0 \text{ m}^2$$

$$A = \frac{4.0 \text{ m} \cdot 10.0 \text{ m}}{2} = 20.0 \text{ m}^2$$

Height (m)	Area (m²)
2.0	4.0
4.0	8.0
6.0	12
8.0	16
10.0	20.0

Table 1.1. Area of a triangle at selected heights.

6. Prepare a graph of area vs. height using the values you computed in the previous step.

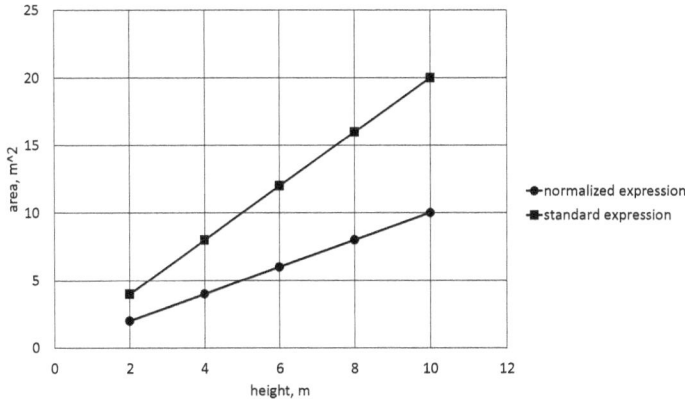

Graph 1.1. Data plot of area vs. height.

7. Compute another table of values for the normalized expression from step 4. Treat the proportional sign as an equals sign for this.

$A \propto h$

$h = 2.0$ m, 4.0 m, 6.0 m, 8.0 m, 10.0 m

$A = 2.0$ m

$A = 4.0$ m

$A = 6.0$ m

$A = 8.0$ m

$A = 10.0$ m

Height (m)	Area (m²)
2.0	2.0
4.0	4.0
6.0	6.0
8.0	8.0
10.0	10.0

Table 1.2. Area of a triangle at selected heights using normalized expression.

8. Graph the normalized equation on the same set of coordinate axes you used for the graph in step 6.

See Graph 1.1.

9. Describe the similarities and differences between the two "curves" on your graph.

Both curves are actually straight lines, the $y = kx$ form. The line for the area of a triangle using the standard expression is steeper than for the normalized expression, but the opposite would be the case if b were less than 1.

10. Answer the main question for this activity.

A varies directly with *h*.

Activity 2. How does the area of a circle vary with its radius, if all else is held constant?

1. What is the relation for area of a circle?

$A = \pi r^2$

2. What are the two key variables that need to be compared in this activity?

A and r

3. Which variable is the independent variable and which is the dependent variable?

A is dependent; r is independent.

4. After combining and normalizing all non-essential variables and constants, what is the expression relating area to radius?

$A \propto r^2$

5. Choose several values for the radius of the circle and calculate the area of the circle for each radius. Enter all these in a table of values.

$A = \pi r^2$

$r = 2.00$ m, 4.00 m, 6.00 m, 8.00 m, 10.0 m

$A = \pi (2.00 \text{ m})^2 = 3.1416 \cdot (2.00 \text{ m})^2 = 12.6 \text{ m}^2$

$A = \pi (4.00 \text{ m})^2 = 3.1416 \cdot (4.00 \text{ m})^2 = 50.3 \text{ m}^2$

$A = \pi (6.00 \text{ m})^2 = 3.1416 \cdot (6.00 \text{ m})^2 = 113 \text{ m}^2$

$A = \pi (8.00 \text{ m})^2 = 3.1416 \cdot (8.00 \text{ m})^2 = 201 \text{ m}^2$

$A = \pi (10.0 \text{ m})^2 = 3.1416 \cdot (10.0 \text{ m})^2 - 314 \text{ m}^2$

Radius (m)	Area (m²)
2.00	12.6
4.00	50.3
6.00	113
8.00	201
10.00	314

Table 2.1. Area of a circle at selected radii using standard expression.

6. Prepare a graph of area vs. radius using the values you computed in the previous step.

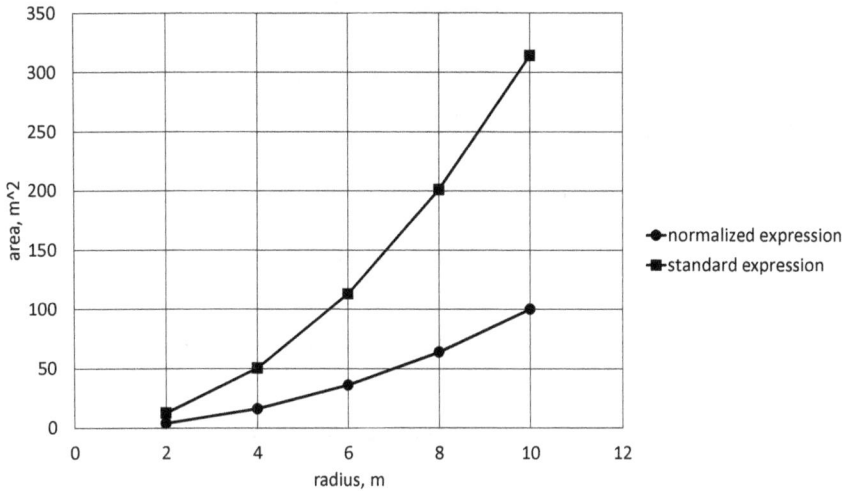

Graph 2.1. Data plot of area vs. radius.

7. Compute another table of values for the normalized expression from step 4. Treat the proportional sign as an equals sign for this.

$A \propto r^2$

$r = 2.00$ m, 4.00 m, 6.00 m, 8.00 m, 10.0 m

$A = (2.00 \text{ m})^2 = 4.00 \text{ m}^2$

$A = (4.00 \text{ m})^2 = 16.0 \text{ m}^2$

$A = (6.00 \text{ m})^2 = 36.0 \text{ m}^2$

$A = (8.00 \text{ m})^2 = 64.0 \text{ m}^2$

$A = (10.0 \text{ m})^2 = 100.0 \text{ m}^2$

Radius (m)	Area (m²)
2.00	4.00
4.00	16.0
6.00	36.0
8.00	64.0
10.00	100.0

Table 2.2. Area of a circle at selected radii using normalized expression.

8. Graph the normalized equation on the same set of coordinate axes you used for the graph in step 6.

See Graph 2.1.

9. Describe the similarities and differences between the two curves on your graph.

Both have the same $y = kx^2$ form, but the curve for the normalized expression is stretched horizontally.

10. Answer the main question for this activity.

A varies as the square of *r*.

Activity 3. How does the volume of a sphere vary with its radius, if all else is held constant?

1. What is the relation for volume of a sphere?

$$V = \frac{4}{3}\pi r^3$$

2. What are the two key variables that need to be compared in this activity?

V and *r*

3. Which variable is the independent variable and which is the dependent variable?

V is dependent; *r* is independent.

4. After combining and normalizing all non-essential variables and constants, what is the expression relating volume to radius?

$$V \propto r^3$$

5. Choose several values for the radius of the sphere and calculate the volume of the sphere for each radius. Enter all of these in a table of values.

$$V = \frac{4}{3}\pi r^3$$

$r = 2.00$ m, 4.00 m, 6.00 m, 8.00 m, 10.0 m

$$V = \frac{4 \cdot 3.1416 \cdot (2.00 \text{ m})^3}{3} = 33.5 \text{ m}^3$$

$$V = \frac{4 \cdot 3.1416 \cdot (4.00 \text{ m})^3}{3} = 268 \text{ m}^3$$

$$V = \frac{4 \cdot 3.1416 \cdot (6.00 \text{ m})^3}{3} = 905 \text{ m}^3$$

$$V = \frac{4 \cdot 3.1416 \cdot (8.00 \text{ m})^3}{3} = 2145 \text{ m}^3$$

$$V = \frac{4 \cdot 3.1416 \cdot (10.0 \text{ m})^3}{3} = 4189 \text{ m}^3$$

Radius (m)	Volume (m³)
2.00	33.5
4.00	268
6.00	905
8.00	2,140
10.0	4,190

Table 3.1. Volume of a sphere at selected radii.

6. Prepare a graph of volume vs. radius using the values you computed in the previous step.

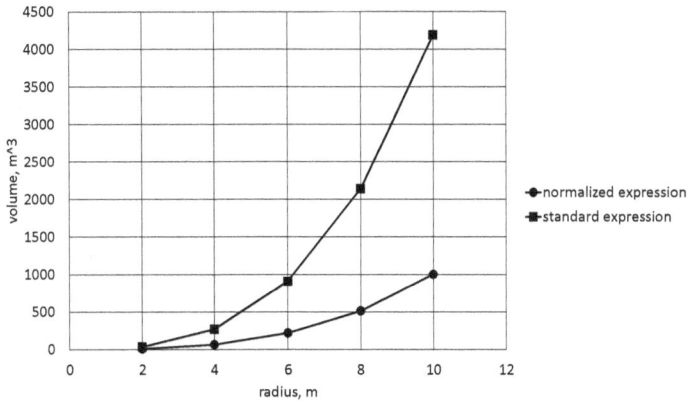

Graph 3.1. Data plot of volume versus radius.

7. Compute another table of values for the normalized expression from step 4. Treat the proportional sign as an equals sign for this.

$$V \propto r^3$$

$r = 2.00 \text{ m}, 4.00 \text{ m}, 6.00 \text{ m}, 8.00 \text{ m}, 10.0 \text{ m}$

$$V = (2.00 \text{ m})^3 = 8.00 \text{ m}^3$$

$$V = (4.00 \text{ m})^3 = 64.0 \text{ m}^3$$

$$V = (6.00 \text{ m})^3 = 216 \text{ m}^3$$

$$V = (8.00 \text{ m})^3 = 512 \text{ m}^3$$

$$V = (10.0 \text{ m})^3 = 1000.0 \text{ m}^3$$

Radius (m)	Volume (m³)
2.00	8.00
4.00	64.0
6.00	216
8.00	512
10.0	1000

Table 3.2. Volume of a sphere at selected radii using normalized expression.

8. Graph the normalized equation on the same set of coordinate axes you used for the graph in step 6.

See Graph 3.1.

9. Describe the similarities and differences between the two "curves" on your graph.

Both curves follow the pattern of $y = kx^3$. The normalized expression shares the same form as the standard expression, but it is stretched horizontally.

10. Answer the main question for this activity.

V varies as the cube of r.

Activity 4. How does the gravitational potential energy of an object vary with its height, if all else is held constant?

1. What are the two key variables that need to be compared in this activity?

E_G and h

2. Which variable is the independent variable and which is the dependent variable?

E_G is dependent and h is independent.

3. After combining and normalizing all non-essential variables and constants, what is the expression relating E_G to height?

$E_G \propto h$

4. Select a value to use for the mass of the object for this activity. Using this value, choose several values for the height of the object and calculate the E_G of the object for each height. Enter all of these in a table of values.

$E_G = mgh$

$m = 2.00$ kg

$g = 9.80 \dfrac{m}{s^2}$

$h = 2.00$ m, 4.00 m, 6.00 m, 8.00 m, 10.0 m

$E_G = 2.00 \text{ kg} \cdot 9.80 \dfrac{m}{s^2} \cdot 2.00 \text{ m} = 39.2 \text{ J}$

$E_G - 2.00 \text{ kg} \cdot 9.80 \dfrac{m}{s^2} \cdot 4.00 \text{ m} = 78.4 \text{ J}$

$E_G = 2.00 \text{ kg} \cdot 9.80 \dfrac{m}{s^2} \cdot 6.00 \text{ m} = 118 \text{ J}$

$E_G = 2.00 \text{ kg} \cdot 9.80 \dfrac{m}{s^2} \cdot 8.00 \text{ m} = 157 \text{ J}$

$E_G = 2.00 \text{ kg} \cdot 9.80 \dfrac{m}{s^2} \cdot 10.0 \text{ m} = 196 \text{ J}$

Height (m)	Gravitational Potential Energy (J)
2.00	39.2
4.00	78.4
6.00	118
8.00	157
10.0	196

Table 4.1. E_G of an object with mass 2.00 kg at selected heights.

5. Prepare a graph of E_G vs. height using the values you computed in the previous step.

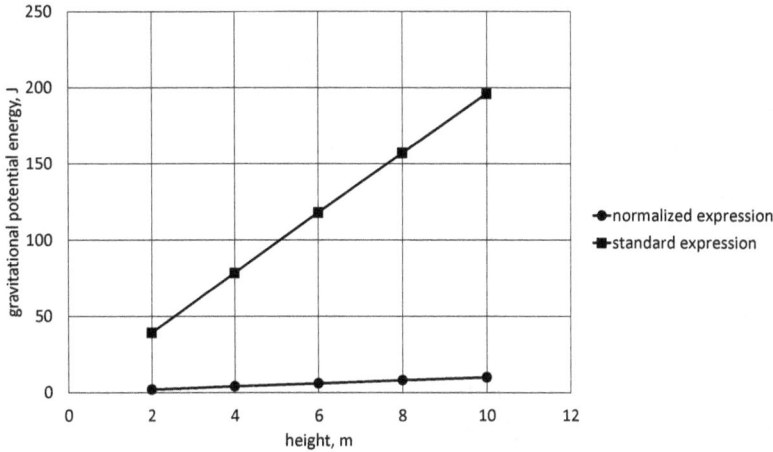

Graph 4.1. Data plot of E_G versus height.

6. Compute another table of values for the normalized expression from step 3. Treat the proportional sign as an equals sign for this.

$E_G \propto h$

$h = 2.00$ m, 4.00 m, 6.00 m, 8.00 m, 10.0 m

$E_G = 2.00$ J

$E_G = 4.00$ J

$E_G = 6.00$ J

$E_G = 8.00$ J

$E_G = 10.0$ J

Height (m)	Gravitational Potential Energy (J)
2.00	2.00
4.00	4.00
6.00	6.00
8.00	8.00
10.0	10.0

Table 4.2. E_G of an object at selected heights using normalized expression.

7. Graph the normalized equation on the same set of coordinate axes you used for the graph in step 5.

See Graph 4.1.

8. Describe the similarities and differences between the two "curves" on your graph.

Both curves are actually straight lines. The normalized expression is stretched horizontally, but both lines follow the pattern of $y = kx$.

9. Answer the main question for this activity.

E_G varies directly with h.

Activity 5. How does the kinetic energy of an object vary with its velocity, if all else is held constant?

---

1. What are the two key variables that need to be compared in this activity?

E_K and v

2. Which variable is the independent variable and which is the dependent variable?

E_K is dependent; v is independent.

3. After combining and normalizing all non-essential variables and constants, what is the expression relating E_K to velocity?

$E_K \propto v^2$

4. Select a value other than 2 kg to use for the mass of the object for this activity. Using this value, choose several values for the velocity of the object and calculate the E_K of the object for each velocity. Enter all these in a table of values.

$E_K = \frac{1}{2}mv^2$

$m = 3.00 \text{ kg}$

$v = 2.00 \dfrac{m}{s},\ 4.00 \dfrac{m}{s},\ 6.00 \dfrac{m}{s},\ 8.00 \dfrac{m}{s},\ 10.0 \dfrac{m}{s}$

$E_K = 0.5 \cdot 3.00 \text{ kg} \cdot \left(2.00\ \dfrac{m}{s} \right)^2 = 6.00 \text{ J}$

$E_K = 0.5 \cdot 3.00 \text{ kg} \cdot \left(4.00\ \dfrac{m}{s} \right)^2 = 24.0 \text{ J}$

$E_K = 0.5 \cdot 3.00 \text{ kg} \cdot \left(6.00\ \dfrac{m}{s} \right)^2 = 54.0 \text{ J}$

$E_K = 0.5 \cdot 3.00 \text{ kg} \cdot \left(8.00\ \dfrac{m}{s} \right)^2 = 96.0 \text{ J}$

$E_K = 0.5 \cdot 3.00 \text{ kg} \cdot \left(10.0\ \dfrac{m}{s} \right)^2 = 150.0 \text{ J}$

Velocity (m/s)	Kinetic Energy (J)
2.0	6.0
4.0	24
6.0	54
8.0	96
10.0	150

Table 5.1. E_K of an object with mass 3.0 kg at selected velocities.

5. Prepare a graph of E_K vs. velocity using the values you computed in the previous step.

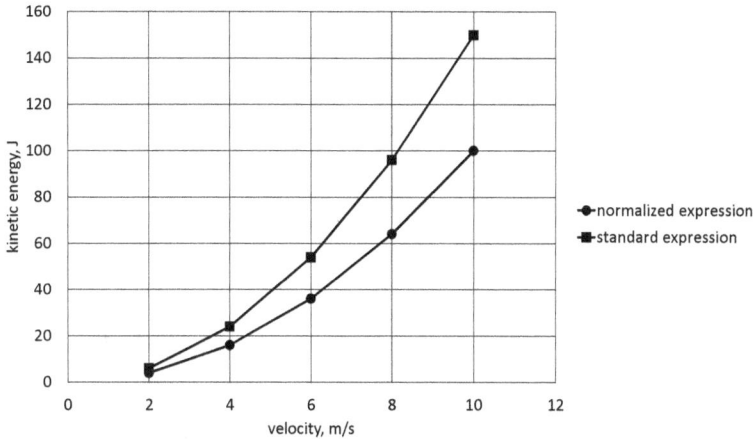

Graph 5.1. Data plot of E_K versus velocity.

6. Compute another table of values for the normalized expression from step 3. Treat the proportional sign as an equals sign for this.

$$E_K \propto v^2$$

$$v = 2.00 \ \frac{m}{s}, \ 4.00 \ \frac{m}{s}, \ 6.00 \ \frac{m}{s}, \ 8.00 \ \frac{m}{s}, \ 10.0 \ \frac{m}{s}$$

$$E_K = \left(2.00 \ \frac{m}{s}\right)^2 = 4.00 \text{ J}$$

$$E_K = \left(4.00 \ \frac{m}{s}\right)^2 = 16.0 \text{ J}$$

$$E_K = \left(6.00 \ \frac{m}{s}\right)^2 = 36.0 \text{ J}$$

$$E_K = \left(8.00 \ \frac{m}{s}\right)^2 = 64.0 \text{ J}$$

$$E_K = \left(10.0 \ \frac{m}{s}\right)^2 = 100.0 \text{ J}$$

Velocity (m/s)	Kinetic Energy (J)
2.0	4.0
4.0	16
6.0	36
8.0	64
10.0	100

Table 5.2. E_K of an object with mass 1.0 kg at selected velocities using normalized expression.

7. Graph the normalized equation on the same set of coordinate axes you used for the graph in step 5.

See Graph 5.1.

8. Describe the similarities and differences between the two curves on your graph.

Both curves share the same form, but the normalized expression is stretched horizontally. The standard expression curve would vary greatly depending on the value chosen for mass.

9. Answer the main question for this activity.

E_K varies as the square of v.

Activity 6. How does pressure under water vary with depth, if all else is held constant?

1. What are the two key variables that need to be compared in this activity?

P and h

2. Which variable is the independent variable and which is the dependent variable?

P is dependent; h is independent.

3. After combining and normalizing all non-essential variables and constants, what is the expression relating pressure to depth?

$P \propto h$

4. Using the value for the density of water given above, choose several values for the depth under water and calculate the pressure for each depth. Enter all these in a table of values.

$P = \rho g h$

$$\rho = 998 \ \frac{kg}{m^3}$$

$$g = 9.80 \ \frac{m}{s^2}$$

$h = 0.020 \text{ m}, 0.040 \text{ m}, 0.060 \text{ m}, 0.080 \text{ m}, 0.100 \text{ m}$

$$P = 998 \ \frac{kg}{m^3} \cdot 9.80 \ \frac{m}{s^2} \cdot 0.020 \text{ m} = 196 \text{ Pa}$$

$$P = 998 \ \frac{kg}{m^3} \cdot 9.80 \ \frac{m}{s^2} \cdot 0.040 \text{ m} = 391 \text{ Pa}$$

$$P = 998 \ \frac{kg}{m^3} \cdot 9.80 \ \frac{m}{s^2} \cdot 0.060 \text{ m} = 587 \text{ Pa}$$

$$P = 998 \ \frac{kg}{m^3} \cdot 9.80 \ \frac{m}{s^2} \cdot 0.080 \text{ m} = 782 \text{ Pa}$$

$$P = 998 \ \frac{kg}{m^3} \cdot 9.80 \ \frac{m}{s^2} \cdot 0.100 \text{ m} = 978 \text{ Pa}$$

Depth (m)	Pressure (Pa)
0.020	196
0.040	391
0.060	587
0.080	782
0.100	978

Table 6.1. Pressure under water at selected depths.

5. Prepare a graph of pressure vs. depth using the values you computed in the previous

step.

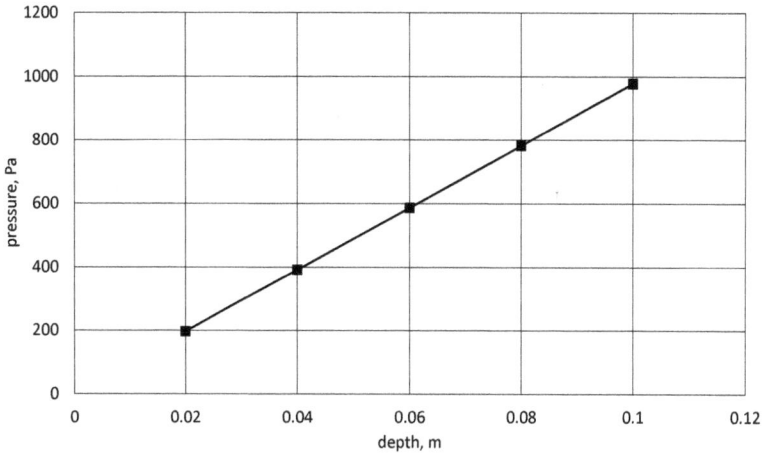

Graph 6.1 Data plot of pressure under water versus depth using standard expression.

6. Compute another table of values for the normalized expression from step 3. Treat the proportional sign as an equals sign for this.

$P \propto h$

$h = 0.020$ m, 0.040 m, 0.060 m, 0.080 m, 0.100 m

$P = 0.020$ Pa

$P = 0.040$ Pa

$P = 0.060$ Pa

$P = 0.080$ Pa

$P = 0.100$ Pa

Depth (m)	Pressure (Pa)
0.020	0.020
0.040	0.040
0.060	0.060
0.080	0.080
0.100	0.100

Table 6.2. Pressure under water at selected depths using normalized expression.

7. Graph the normalized equation on a separate set of coordinate axes you used for the

graph in step 5.

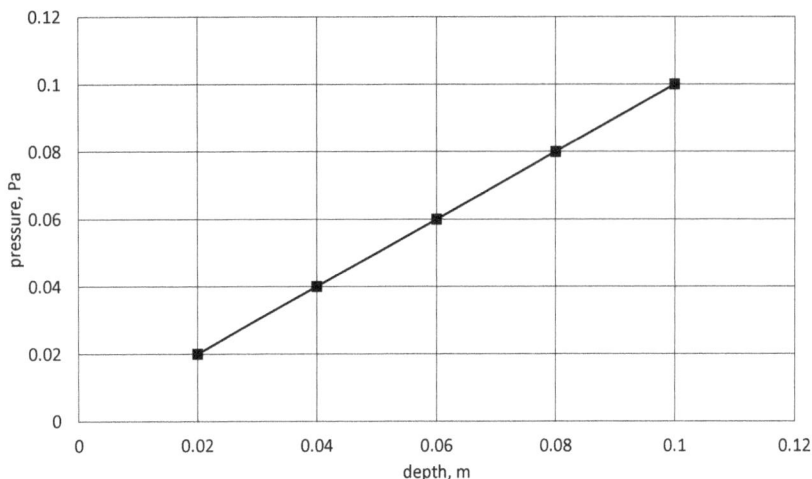

Graph 6.2. Data plot of pressure versus depth using normalized expression.

8. Describe the similarities and differences between the curves in the two graphs.

Both are linear, following the $y = kx$ form, but the standard expression is much, much steeper due to the high density of water. It would be virtually meaningless to graph them on the same set of axes, since the normalized expression would almost look like it was a horizontal line.

9. Answer the main question for this activity.

Pressure under water varies directly with the depth.

Activity 7. How does the force of gravitational attraction vary with the distance between the centers of two objects?

1. What are the two key variables that need to be compared in this activity?

F and d

2. Which variable is the independent variable and which is the dependent variable?

F is dependent; d is independent.

3. After combining and normalizing all non-essential variables and constants, what is the expression relating the gravitational force between two objects to the distance between them?

$$F \propto \frac{1}{d^2}$$

4. Select values to use for the two masses of the objects for this activity. Using these values, choose several values for the distance between the objects and calculate the force of

attraction for each distance. Enter all these in a table of values.

$$F = G\frac{m_1 m_2}{d^2}$$

$$G = 6.67 \times 10^{-11} \; \frac{N \cdot m^2}{kg^2}$$

$m_1 = 2.5$ kg

$m_2 = 3.0$ kg

$d = 2.0$ m, 4.0 m, 6.0 m, 8.0 m, 10.0 m

$$F = 6.67 \times 10^{-11} \; \frac{N \cdot m^2}{kg^2} \cdot \frac{2.5 \; kg \cdot 3.0 \; kg}{(2.0 \; m)^2} = 1.3 \times 10^{-10} \; N = 130 \times 10^{-12} \; N$$

$$F = 6.67 \times 10^{-11} \; \frac{N \cdot m^2}{kg^2} \cdot \frac{2.5 \; kg \cdot 3.0 \; kg}{(4.0 \; m)^2} = 3.1 \times 10^{-11} \; N = 31 \times 10^{-12} \; N$$

$$F = 6.67 \times 10^{-11} \; \frac{N \cdot m^2}{kg^2} \cdot \frac{2.5 \; kg \cdot 3.0 \; kg}{(6.0 \; m)^2} = 1.4 \times 10^{-11} \; N = 14 \times 10^{-12} \; N$$

$$F = 6.67 \times 10^{-11} \; \frac{N \cdot m^2}{kg^2} \cdot \frac{2.5 \; kg \cdot 3.0 \; kg}{(8.0 \; m)^2} = 7.9 \times 10^{-12} \; N$$

$$F = 6.67 \times 10^{-11} \; \frac{N \cdot m^2}{kg^2} \cdot \frac{2.5 \; kg \cdot 3.0 \; kg}{(10.0 \; m)^2} = 5.0 \times 10^{-12} \; N$$

Distance (m)	Gravitational force (N)
2.0	130×10^{-12}
4.0	31×10^{-12}
6.0	14×10^{-12}
8.0	7.9×10^{-12}
10.0	5.0×10^{-12}

Table 7.1. Gravitational force between two objects of mass 2.5 and 3.0 kg at selected distances. (Note: To facilitate graphing, all values are expressed using the same power of 10.)

5. Prepare a graph of force vs. distance using the values you computed in the previous step.

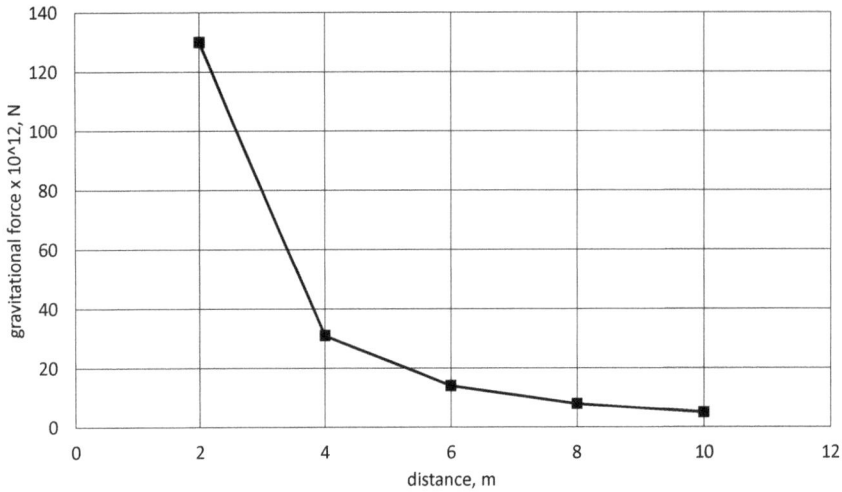

Graph 7.1. Data plot of gravitational force versus distance between two objects of mass 2.5 and 3.0 kg.

6. Compute another table of values for the normalized expression from step 3. Treat the proportional sign as an equals sign for this.

$$F \propto \frac{1}{d^2}$$

$d = 2.0$ m, 4.0 m, 6.0 m, 8.0 m, 10.0 m

$$F = \frac{1}{(2.0 \text{ m})^2} = 0.25 \text{ N}$$

$$F = \frac{1}{(4.0 \text{ m})^2} = 0.0625 \text{ N}$$

$$F = \frac{1}{(6.0 \text{ m})^2} = 0.0278 \text{ N}$$

$$F = \frac{1}{(8.0 \text{ m})^2} = 0.0156 \text{ N}$$

$$F = \frac{1}{(10.0 \text{ m})^2} = 0.0100 \text{ N}$$

Distance (m)	Gravitational force (N)
2.0	0.25
4.0	0.0625
6.0	0.0278
8.0	0.0156
10.0	0.0100

Table 7.2. Gravitational force between two objects at selected distances using normalized expression.

7. Graph the normalized equation on a separate set of coordinate axes you used for the

graph in step 5.

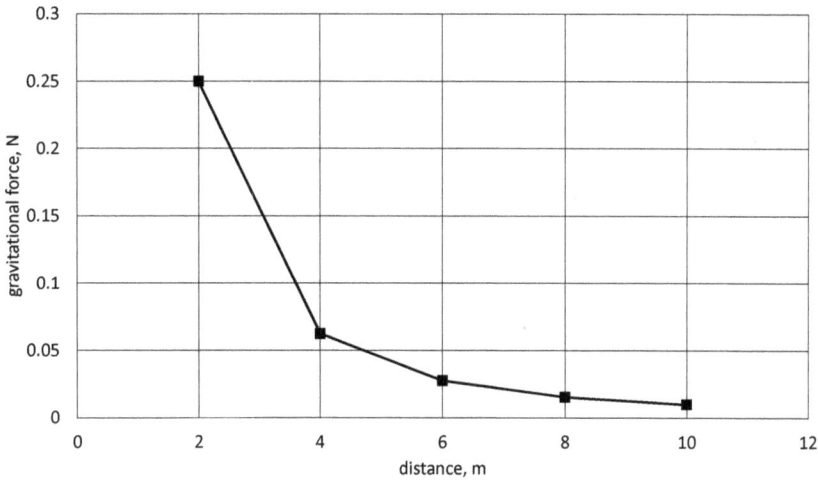

Graph 7.2. Gravitational force versus distance between two objects using normalized expression.

8. Describe the similarities and differences between the two curves on your graph.

The curves are almost identical, and follow the $y = k/x^2$ form. However, the scales are vastly different.

9. Answer the main question for this activity.

Force of gravitational attraction varies as the inverse square of distance between two objects.

Activity 8. How does the volume of a gas vary with its temperature?

The values in the first, third, and fifth columns presented below were taken from the "Demo Data" found in Favorite Experiments, Part 2, Demo #3 on Variation and Proportion. Details for the book can be found at novarescienceandmath.com.

First, enter the data from the demo in your table.

T_C (°C)	T_K (K)	V_{buret} (mL = cm³)	ΔV (cm³)	$V_{experimental}$ (cm³)	$V_{predicted}$
3.7		24.7			
9.8		24.6			
15.0		24.5			
18.4		24.4			
23.5		24.3			
28.8		24.2			
33.8		24.2			
37.0		24.0			

Table 8.1. Initial table for Charles' law activity with demo data.

Next, convert all the temperature values from degrees Celsius to kelvins.

$T_K = T_C + 273.2$

$T_K = 3.7°C + 273.2 = 276.9 \text{ K}$

$T_K = 9.8°C + 273.2 = 283.0 \text{ K}$

$T_K = 15.0°C + 273.2 = 288.2 \text{ K}$

$T_K = 18.4°C + 273.2 = 291.6 \text{ K}$

$T_K = 23.5°C + 273.2 = 296.7 \text{ K}$

$T_K = 28.8°C + 273.2 = 302.0 \text{ K}$

$T_K = 33.8°C + 273.2 = 307.0 \text{ K}$

$T_K = 37.0°C + 273.2 = 310.2 \text{ K}$

Enter all the temperatures in kelvins in the table. Next, we obtain the values for ΔV.

$\Delta V = \left(\text{first reading of } V_{buret} \right) - V_{buret}$

$\Delta V = 24.7 \text{ cm}^3 - 24.7 \text{ cm}^3 = 0 \text{ cm}^3$

$\Delta V = 24.7 \text{ cm}^3 - 24.6 \text{ cm}^3 = 0.1 \text{ cm}^3$

$\Delta V = 24.7 \text{ cm}^3 - 24.5 \text{ cm}^3 = 0.2 \text{ cm}^3$

$\Delta V = 24.7 \text{ cm}^3 - 24.4 \text{ cm}^3 = 0.3 \text{ cm}^3$

$\Delta V = 24.7 \text{ cm}^3 - 24.3 \text{ cm}^3 = 0.4 \text{ cm}^3$

$\Delta V = 24.7 \text{ cm}^3 - 24.2 \text{ cm}^3 = 0.5 \text{ cm}^3$

$\Delta V = 24.7 \text{ cm}^3 - 24.1 \text{ cm}^3 = 0.6 \text{ cm}^3$

$\Delta V = 24.7 \text{ cm}^3 - 24.0 \text{ cm}^3 = 0.7 \text{ cm}^3$

Next we determine the value of the initial air volume in the buret, V_i.

$$V_{\text{experimental}} = V_i + \Delta V$$

$$V_i = ?$$

$$V = \frac{V_i}{T_i} T$$

$$V_i + \Delta V = \frac{V_i}{T_i} T$$

$$V_i - \frac{V_i}{T_i} T = -\Delta V$$

$$\frac{V_i}{T_i} T - V_i = \Delta V$$

$$V_i \left(\frac{T}{T_i} - 1 \right) = \Delta V$$

$$V_i = \frac{\Delta V}{\dfrac{T}{T_i} - 1}$$

$$T_K = T_C + 273.2 = 37°C + 273.2 = 310.2 \text{ K}$$

$$V_i = \frac{0.7 \text{ cm}^3}{\left(\dfrac{310.2 \text{ K}}{276.9 \text{ K}} - 1 \right)} = 5.82 \text{ cm}^3$$

Finally, now that we have V_i, we use it with T_i and the values of T to calculate the predicted value of the volume for each temperature.

$$V_{\text{predicted}} = \frac{V_i}{T_i} T$$

$$V_{\text{predicted}} = \frac{5.82 \text{ cm}^3}{276.9 \text{ K}} \cdot 283.0 \text{ K} = 5.95 \text{ cm}^3$$

$$V_{\text{predicted}} = \frac{5.82 \text{ cm}^3}{276.9 \text{ K}} \cdot 288.2 \text{ K} = 6.06 \text{ cm}^3$$

$$V_{\text{predicted}} = \frac{5.82 \text{ cm}^3}{276.9 \text{ K}} \cdot 291.6 \text{ K} = 6.13 \text{ cm}^3$$

$$V_{\text{predicted}} = \frac{5.82 \text{ cm}^3}{276.9 \text{ K}} \cdot 296.7 \text{ K} = 6.24 \text{ cm}^3$$

$$V_{\text{predicted}} = \frac{5.82 \text{ cm}^3}{276.9 \text{ K}} \cdot 302.0 \text{ K} = 6.35 \text{ cm}^3$$

$$V_{\text{predicted}} = \frac{5.82 \text{ cm}^3}{276.9 \text{ K}} \cdot 307.0 \text{ K} = 6.45 \text{ cm}^3$$

$$V_{\text{predicted}} = \frac{5.82 \text{ cm}^3}{276.9 \text{ K}} \cdot 310.2 \text{ K} = 6.52 \text{ cm}^3$$

T_C (°C)	T_K (K)	V_{buret} (mL = cm³)	ΔV (cm³)	$V_{experimental}$ (cm³)	$V_{predicted}$
3.7	T_i = 276.9	24.7	0	V_i = 5.82	V_i = 5.82
9.8	283.0	24.6	0.1	5.92	5.95
15.0	288.2	24.5	0.2	6.02	6.06
18.4	291.6	24.4	0.3	6.12	6.13
23.5	296.7	24.3	0.4	6.22	6.24
28.8	302.0	24.2	0.5	6.32	6.35
33.8	307.0	24.2	0.6	6.42	6.45
37.0	310.2	24.0	0.7	6.52	6.52

Table 8.2. Complete table for Charles' law activity.

The result of our work is a graph of volume vs. temperature, showing both predicted and experimental values of volume for each value of temperature.

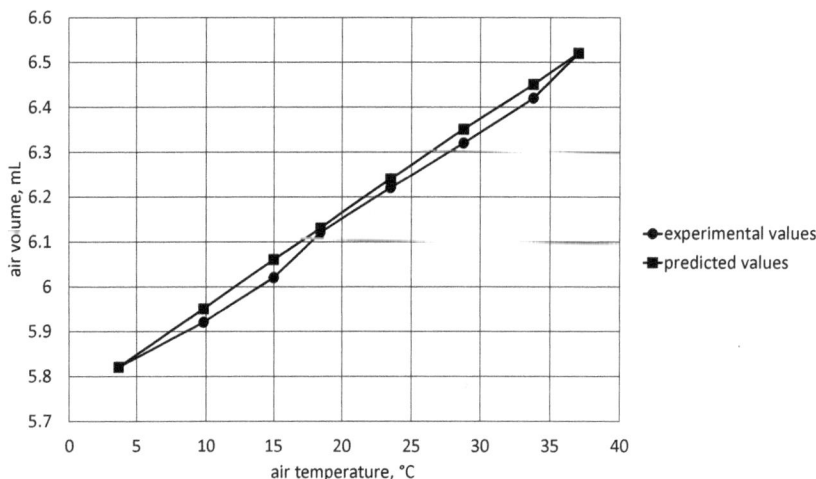

Graph 8.1. Volume vs. temperature for Charles' law demonstration.

Activity 9. How does the volume of a gas vary with its pressure?

1. What are the two key variables that need to be compared in this activity?

V and P

2. Which variable is the independent variable and which is the dependent variable?

V is dependent; P is independent.

3. After combining and normalizing all non-essential variables and constants, what is the expression relating volume to pressure?

$$V \propto \frac{1}{P}$$

4. How are these variables related to one another?

V varies inversely with P.

5. Your job is to use Boyle's law to calculate the volume of a balloon at various pressures from atmospheric pressure at the surface of the water down to the pressure at a depth of 50.0 m. So your lowest pressure value is at the surface; the highest pressure value is the pressure at 50.0 m deep. You can pick several pressures in between and calculate the volume for them as well.

Graph the changes in the volume as the pressure is increased from atmospheric pressure at the surface to pressure at final depth.

First, verify the given initial volume:

$$r = \frac{9.39 \text{ in}}{2} \cdot \frac{2.54 \text{ cm}}{\text{in}} \cdot \frac{1 \text{ m}}{100 \text{ cm}} = 0.1193 \text{ m}$$

$$V = \frac{4}{3}\pi r^3 = \frac{4 \cdot 3.14159 \cdot (0.1193 \text{ m})^3}{3} = 0.00711 \text{ m}^3$$

Next, determine the pressure at a depth of 50.0 m:

$$P = 101,325 \text{ Pa} + \rho g h$$

$$P = 101,325 \text{ Pa} + 998 \frac{\text{kg}}{\text{m}^3} \cdot 9.80 \frac{\text{m}}{\text{s}^2} \cdot 50.0 \text{ m} = 590,345 \text{ Pa} \cdot \frac{1 \text{ kPa}}{1000 \text{ Pa}} = 590.3 \text{ kPa}$$

Next, calculate the volumes for several intermediate pressures:

$$V = \frac{V_i P_i}{P}$$

$P_i = 101.3 \text{ kPa}$

$V_i = 0.00711 \text{ m}^3$

$P = 101.3 \text{ kPa}, 200 \text{ kPa}, 300 \text{ kPa}, 400 \text{ kPa}, 500 \text{ kPa}, 590.3 \text{ kPa}$

$$V = \frac{0.00711 \text{ m}^3 \cdot 101.3 \text{ kPa}}{200 \text{ kPa}} = 0.00360 \text{ m}^3$$

$$V = \frac{0.00711 \text{ m}^3 \cdot 101.3 \text{ kPa}}{300 \text{ kPa}} = 0.00240 \text{ m}^3$$

$$V = \frac{0.00711 \text{ m}^3 \cdot 101.3 \text{ kPa}}{400 \text{ kPa}} = 0.00180 \text{ m}^3$$

$$V = \frac{0.00711 \text{ m}^3 \cdot 101.3 \text{ kPa}}{500 \text{ kPa}} = 0.00144 \text{ m}^3$$

$$V = \frac{0.00711 \text{ m}^3 \cdot 101.3 \text{ kPa}}{590.3 \text{ kPa}} = 0.00122 \text{ m}^3$$

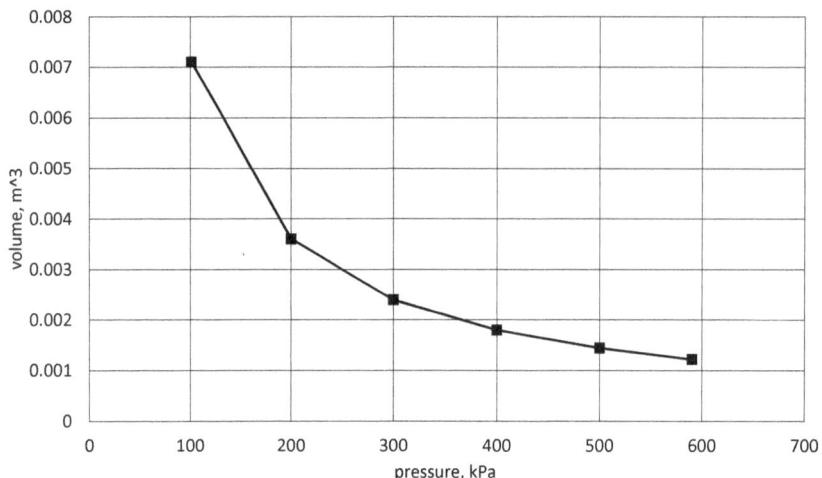

Graph 9.1. Data plot of volume versus pressure.

6. Find out how big this balloon is at this depth. Take the final volume you calculated for a depth of 50.0 m and use the equation below to calculate the radius of the balloon at 50.0 m depth. Then double it to get the diameter. Give your final result in inches.

$$r = \left(\frac{3V}{4\pi} \right)^{1/3}$$

$$r = \left(\frac{3 \cdot 0.00122 \ m^3}{4 \cdot 3.14159} \right)^{1/3} = 0.06629 \ m \cdot \frac{1 \ ft}{0.3048 \ m} \cdot \frac{12 \ in}{1 \ ft} = 2.61 \ in$$

$$D = 2r = 2 \cdot 2.61 \ in = 5.22 \ in$$

Chapter 5

1.

$m = 1.00 \times 10^5 \text{ kg}$

$h = 240 \text{ ft} \cdot \dfrac{0.3048 \text{ m}}{1 \text{ ft}} = 73.15 \text{ m}$

$g = 9.80 \dfrac{\text{m}}{\text{s}^2}$

$E_G = ?$

$E_G = mgh = 1.00 \times 10^5 \text{ kg} \cdot 9.80 \dfrac{\text{m}}{\text{s}^2} \cdot 73.15 \text{ m} = 72{,}000{,}000 \text{ J}$

2.

$m = 25 \text{ g} \cdot \dfrac{1 \text{ kg}}{1000 \text{ g}} = 0.025 \text{ kg}$

$v = 556 \dfrac{\text{ft}}{\text{s}} \cdot \dfrac{0.3048 \text{ m}}{1 \text{ ft}} = 169.5 \dfrac{\text{m}}{\text{s}}$

$E_K = ?$

$E_K = \dfrac{1}{2} mv^2 = \dfrac{1}{2} \cdot 0.025 \text{ kg} \cdot \left(169.5 \dfrac{\text{m}}{\text{s}} \right)^2 = 360 \text{ J}$

3.

$d = 75 \text{ cm} \cdot \dfrac{1 \text{ m}}{100 \text{ cm}} = 0.75 \text{ m}$

$m = 12{,}500 \text{ g} \cdot \dfrac{1 \text{ kg}}{1000 \text{ g}} = 12.5 \text{ kg}$

$W = ?$

$W = Fd$

$F_w = mg = 12.5 \text{ kg} \cdot 9.80 \dfrac{\text{m}}{\text{s}^2} = 122.5 \text{ kg}$

$W = 122.5 \text{ kg} \cdot 0.75 \text{ m} = 92 \text{ J}$

4.

$m = 12.5$ kg

$h = 0.75$ m

$$E_G = mgh = 0.75 \text{ m} \cdot 9.80 \ \frac{\text{m}}{\text{s}^2} \cdot 12.5 \text{ kg} = 92 \text{ J}$$

5.

$m = 12.5$ kg

$h_i = 0.75$ m

$h_f = 0$

$v_i = 0$

$v_f = ?$

$E_{Gi} + E_{Ki} = E_{Gf} + E_{Kf}$

$E_{Kf} = E_{Gi} + E_{Ki} - E_{Gf} = 92 \text{ J} + 0 - 0 = 92 \text{ J}$

$$v_f = \sqrt{\frac{2E_{Kf}}{m}} = \sqrt{\frac{2 \cdot 92 \text{ J}}{12.5 \text{ kg}}} = 3.8 \ \frac{\text{m}}{\text{s}}$$

6.a.

$$m = 255.8 \text{ g} \cdot \frac{1 \text{ kg}}{1000 \text{ g}} = 0.2558 \text{ kg}$$

$$h_i = 10.4 \text{ ft} \cdot \frac{0.3048 \text{ m}}{1 \text{ ft}} = 3.1699 \text{ m}$$

$E_{Gi} = ?$

$$E_{Gi} = mgh_i = 0.2558 \text{ kg} \cdot 9.80 \ \frac{\text{m}}{\text{s}^2} \cdot 3.1699 \text{ m} = 7.95 \text{ J}$$

6.b.

$h_f = 0$

$v_i = 0$

$v_f = ?$

$E_{Gi} + E_{Ki} = E_{Gf} + E_{Kf}$

$E_{Ki} = 0$

$E_{Gf} = 0$

$E_{Kf} = E_{Gi} + E_{Ki} - E_{Gf} = 7.95 \text{ J} + 0 \text{ J} - 0 \text{ J} = 7.95 \text{ J}$

$$v_f = \sqrt{\frac{2E_{Kf}}{m}} = \sqrt{\frac{2 \cdot 7.95 \text{ J}}{0.2558 \text{ kg}}} = 7.88 \ \frac{\text{m}}{\text{s}}$$

Energy Calculations Set 1

1.

$m = 1.31 \times 10^3 \text{ kg}$

$h = 177.44 \text{ ft} \cdot \dfrac{0.3048 \text{ m}}{1 \text{ ft}} = 54.084 \text{ m}$

$E_G = ?$

$E_G = mgh = 1.31 \times 10^3 \text{ kg} \cdot 9.80 \ \dfrac{\text{m}}{\text{s}^2} \cdot 54.084 \text{ m} = 694{,}000 \text{ J}$

2.

$m = 2345 \text{ kg}$

$v = 31 \ \dfrac{\text{mi}}{\text{hr}} \cdot \dfrac{5280 \text{ ft}}{1 \text{ mi}} \cdot \dfrac{0.3048 \text{ m}}{1 \text{ ft}} \cdot \dfrac{1 \text{ hr}}{3600 \text{ s}} = 13.858 \ \dfrac{\text{m}}{\text{s}}$

$E_K = \tfrac{1}{2}mv^2 = 0.5 \cdot 2345 \text{ kg} \cdot \left(13.858 \ \dfrac{\text{m}}{\text{s}}\right)^2 = 230{,}000 \text{ J}$

3.

$$d = 61.7 \text{ cm} \cdot \frac{1 \text{ m}}{100 \text{ cm}} = 0.617 \text{ m}$$

$m = 17.5 \text{ kg}$

$W = ?$

$$a = \frac{F}{m}$$

$$F_w = mg = 17.5 \text{ kg} \cdot 9.80 \, \frac{\text{m}}{\text{s}^2} = 171.5 \text{ N}$$

$$W = F_w d = 171.5 \text{ N} \cdot 0.617 \text{ m} = 106 \text{ J}$$

4.

$$h = 61.7 \text{ cm} \cdot \frac{1 \text{ m}}{100 \text{ cm}} = 0.617 \text{ m}$$

$m = 17.5 \text{ kg}$

$E_G = ?$

$$E_G = mgh = 17.5 \text{ kg} \cdot 9.80 \, \frac{\text{m}}{\text{s}^2} \cdot 0.617 \text{ m} = 106 \text{ J}$$

5.

$m = 17.5 \text{ kg}$

$h_f = 0.617 \text{ m}$

$h_i = 0$

$v_i = 0$

$v_f = ?$

$$E_{Gi} + E_{Ki} = E_{Gf} + E_{Kf}$$

$$E_{Kf} = E_{Gi} + E_{Ki} - E_{Gf} = 106 \text{ J} + 0 \text{ J} - 0 \text{ J} = 106 \text{ J}$$

$$v_f = \sqrt{\frac{2E_{Kf}}{m}} = \sqrt{\frac{2 \cdot 106 \text{ J}}{17.5 \text{ kg}}} = 3.48 \, \frac{\text{m}}{\text{s}}$$

6.

$$m = 122 \text{ g} \cdot \frac{1 \text{ kg}}{1000 \text{ g}} = 0.122 \text{ kg}$$

$$v_i = 13.75 \frac{\text{m}}{\text{s}}$$

$$v_f = 0$$

$$h_i = 0$$

$$h_f = ?$$

$$E_{Ki} = \tfrac{1}{2}mv^2 = 0.5 \cdot 0.122 \text{ kg} \cdot \left(13.75 \frac{\text{m}}{\text{s}}\right)^2 = 11.53 \text{ J}$$

$$E_{Gi} + E_{Ki} = E_{Gf} + E_{Kf}$$

$$E_{Gf} = E_{Gi} + E_{Ki} - E_{Kf} = 0 + 11.53 \text{ J} - 0 = 11.53 \text{ J}$$

$$E_{Gf} = mgh_f$$

$$h_f = \frac{E_{Gf}}{mg} = \frac{11.53 \text{ J}}{0.122 \text{ kg} \cdot 9.8 \frac{\text{m}}{\text{s}^2}} = 9.65 \text{ m}$$

7.

$$m = 325 \text{ g} \cdot \frac{1 \text{ kg}}{1000 \text{ g}} = 0.325 \text{ kg}$$

$$h_i = 36.1 \text{ m}$$

$$h_f = 0$$

$$v_i = 0$$

$$v_f = ?$$

$$E_{Gi} = mgh_i = 0.325 \text{ kg} \cdot 9.80 \frac{\text{m}}{\text{s}^2} \cdot 36.1 \text{ m} = 114.98 \text{ J}$$

$$E_{Gi} + E_{Ki} = E_{Gf} + E_{Kf}$$

$$E_{Kf} = E_{Gi} + E_{Ki} - E_{Gf} = 114.98 \text{ J} + 0 - 0$$

$$E_{Kf} = 114.98 \text{ J}$$

$$E_{Kf} = \frac{1}{2}mv^2$$

$$v_f = \sqrt{\frac{2E_{Kf}}{m}} = \sqrt{\frac{2 \cdot 114.98 \text{ J}}{0.325 \text{ kg}}} = 26.6 \frac{\text{m}}{\text{s}}$$

8.

$F = 735$ N

$d = 26$ m

$W = ?$

$W = Fd = 735$ N$\cdot 26$ m $= 19,000$ J

Energy Calculations Set 2

1.a.

$F_w = 20 \cdot 80.0 \text{ lb} \cdot \dfrac{4.45 \text{ N}}{1 \text{ lb}} = 7120 \text{ N}$

$h = 8.5$ m

$g = 9.80 \dfrac{\text{m}}{\text{s}^2}$

$m = ?$

$F_w = mg$

$m = \dfrac{F_w}{g}$

$m = \dfrac{7120 \text{ N}}{9.80 \dfrac{\text{m}}{\text{s}^2}} = 727 \text{ kg}$

1.b

$F = 7120$ N

$d = 8.5$ m

$W = Fd = 7120$ N$\cdot 8.5$ m

$W = 6.0 \times 10^4$ J

1.c.

$m = 727$ kg

$g = 9.80 \dfrac{\text{m}}{\text{s}^2}$

$h = 8.5$ m

$E_G = mgh$

$E_G = 727 \text{ kg} \cdot 9.80 \dfrac{\text{m}}{\text{s}^2} \cdot 8.5 \text{ m} = 6.0 \times 10^4 \text{ J}$

1.d.i.

$m = 727 \text{ kg}$

$v_i = 0$

$E_{Ki} = ?$

$$E_{Ki} = \frac{1}{2}mv_i^2 = \frac{1}{2} \cdot 727 \text{ kg} \cdot 0^2 = 0 \text{ J}$$

1.d.ii.

$E_{Gi} = 6.0 \times 10^4 \text{ J}$

$E_{Gf} = 0$

$E_{Kf} = ?$

$E_{Gi} + E_{Ki} = E_{Gf} + E_{Kf}$

$E_{Kf} = E_{Gi} + E_{Ki} - E_{Gf} = 6.0 \times 10^4 \text{ J} + 0 - 0 = 6.0 \times 10^4 \text{ J}$

1.e.

$m = 727 \text{ kg}$

$E_{Kf} = 6.0 \times 10^4 \text{ J}$

$v_f = ?$

$E_{Kf} = \frac{1}{2}mv_f^2$

$$v_f = \sqrt{\frac{2E_{Kf}}{m}} = \sqrt{\frac{2 \cdot 6.0 \times 10^4 \text{ J}}{727 \text{ kg}}} = 13 \frac{\text{m}}{\text{s}}$$

2.a.

$$F_w = 3{,}193 \text{ lb} \cdot \frac{4.45 \text{ N}}{1 \text{ lb}} = 14{,}209 \text{ N}$$

$m = ?$

$F_w = mg$

$$m = \frac{F_w}{g}$$

$$m = \frac{14{,}209 \text{ N}}{9.80 \ \dfrac{\text{m}}{\text{s}^2}} = 1450 \text{ kg}$$

$h = 16 \text{ m}$

$$E_G = mgh = 1450 \text{ kg} \cdot 9.80 \ \frac{\text{m}}{\text{s}^2} \cdot 16 \text{ m} = 227{,}360 \text{ J}$$

$$E_G = 230{,}000 \text{ J}$$

2.c.

$E_{Kf} = ?$

$E_{G_i} + E_{Ki} = E_{Gf} + E_{Kf}$

$E_{Kf} = E_{G_i} + E_{Ki} - E_{Gf}$

$E_{Kf} = 230{,}000 \text{ J} + 0 - 0 = 230{,}000 \text{ J}$

2.d.

$E_{Kf} = 230{,}000 \text{ J}$

$m = 1450 \text{ kg}$

$v_f = ?$

$$v_f = \sqrt{\frac{2E_{Kf}}{m}} = \sqrt{\frac{2 \cdot 230{,}000 \text{ J}}{1450 \text{ kg}}} = 18 \ \frac{\text{m}}{\text{s}}$$

3.a.

$m = 1450 \text{ kg}$

$h = 8 \text{ m}$

$E_{G8(at \ h = 8)} = ?$

$$E_{G8} = mgh = 1450 \text{ kg} \cdot 9.80 \ \frac{\text{m}}{\text{s}^2} \cdot 8 \text{ m} = 113{,}680 \text{ J}$$

$$E_{G8} = 114{,}000 \text{ J}$$

3.b.

$m = 1450$ kg

$h = 8$ m

$E_{G8} = 113,680$ J

$E_{K8} = ?$

$E_{G8} + E_{K8} = E_{Gf} + E_{Kf}$

$E_{K8} = E_{Gf} + E_{Kf} - E_{G8} = 0 + 227,360 \text{ J} - 113,680 \text{ J} = 114,000 \text{ J}$

3.c.

$E_{K8} = 114,000$ J

$v_8 = ?$

$m = 1450$ kg

$$v_8 = \sqrt{\frac{2E_{K8}}{m}} = \sqrt{\frac{2 \cdot 114,000 \text{ J}}{1450 \text{ kg}}} = 13 \; \frac{\text{m}}{\text{s}}$$

4.a. 1st calc

$m = 1450$ kg

$g = 9.80 \; \dfrac{\text{m}}{\text{s}^2}$

$h = 16.0$ m

$E_{G16(at \; h = 16m)} = ?$

$E_{G16} = mgh = 1450 \text{ kg} \cdot 9.80 \; \dfrac{\text{m}}{\text{s}^2} \cdot 16.0 \text{ m} = 227,360 \text{ J}$

$E_{K16(at \; h = 16m)} = 0$

$v_{16} = 0$

4.a. 2nd calc

$m = 1450$ kg

$g = 9.80 \dfrac{m}{s^2}$

$h = 14.0$ m

$E_{G14(at\ h\ =\ 14m)} = ?$

$E_{G14} = mgh = 1450$ kg $\cdot 9.80 \dfrac{m}{s^2} \cdot 14.0$ m $= 199{,}000$ J

$E_{K14(at\ h\ =\ 14m)} = ?$

$E_{G16} + E_{K16} = E_{G14} + E_{K14}$

$E_{K14} = E_{G16} + E_{K16} - E_{G14} = 227{,}360$ J $+ 0 - 198{,}940$ J

$E_{K14} = 28{,}400$ J

$v_{14} = ?$

$v_{14} = \sqrt{\dfrac{2E_{K14}}{m}} = \sqrt{\dfrac{2 \cdot 28{,}420\ \text{J}}{1450\ \text{kg}}} = 6.26 \dfrac{m}{s}$

4.a. 3rd calc

$m = 1450$ kg

$g = 9.80 \dfrac{m}{s^2}$

$h = 12.0$ m

$E_{G12} = ?$

$E_{G12} = mgh = 1450$ kg $\cdot 9.80 \dfrac{m}{s^2} \cdot 12.0$ m $= 170{,}520$ J

$E_{K12} = ?$

$E_{G16} + E_{K16} = E_{G12} + E_{K12}$

$E_{K12} = E_{G16} + E_{K16} - E_{G12} = 227{,}360$ J $+ 0 - 170{,}520$ J

$E_{K12} = 56{,}800$ J

$v_{12} = ?$

$v_{12} = \sqrt{\dfrac{2E_{K12}}{m}} = \sqrt{\dfrac{2 \cdot 56{,}840\ \text{J}}{1450\ \text{kg}}} = 8.85 \dfrac{m}{s}$

4.a. 4th calc

$m = 1450 \text{ kg}$

$g = 9.80 \ \dfrac{\text{m}}{\text{s}^2}$

$h = 10.0 \text{ m}$

$E_{G10} = ?$

$E_{G10} = mgh = 1450 \text{ kg} \cdot 9.80 \ \dfrac{\text{m}}{\text{s}^2} \cdot 10.0 \text{ m} = 142{,}000 \text{ J}$

$E_{K10} = ?$

$E_{G16} + E_{K16} = E_{G10} + E_{K10}$

$E_{K10} = E_{G16} + E_{K16} - E_{G10} = 227{,}360 \text{ J} + 0 - 142{,}100 \text{ J}$

$E_{K10} = 85{,}300 \text{ J}$

$v_{10} = ?$

$v_{10} = \sqrt{\dfrac{2 E_{K10}}{m}} = \sqrt{\dfrac{2 \cdot 85{,}260 \text{ J}}{1450 \text{ kg}}} = 10.8 \ \dfrac{\text{m}}{\text{s}}$

4.a. 5th calc

$m = 1450 \text{ kg}$

$g = 9.80 \ \dfrac{\text{m}}{\text{s}^2}$

$h = 8.00 \text{ m}$

$E_{G8} = ?$

$E_{G8} = mgh = 1450 \text{ kg} \cdot 9.80 \ \dfrac{\text{m}}{\text{s}^2} \cdot 8.00 \text{ m} = 113{,}700 \text{ J}$

$E_{K8} = ?$

$E_{G16} + E_{K16} = E_{G8} + E_{K8}$

$E_{K8} = E_{G16} + E_{K16} - E_{G8} = 227{,}360 \text{ J} + 0 - 113{,}700 \text{ J}$

$E_{K8} = 113{,}700 \text{ J}$

$v_8 = ?$

$v_8 = \sqrt{\dfrac{2 E_{K8}}{m}} = \sqrt{\dfrac{2 \cdot 113{,}700 \text{ J}}{1450 \text{ kg}}} = 12.5 \ \dfrac{\text{m}}{\text{s}}$

4.a. 6th calc

$m = 1450$ kg

$g = 9.80 \dfrac{m}{s^2}$

$h = 6.0$ m

$E_{G6} = ?$

$E_{G6} = mgh = 1450 \text{ kg} \cdot 9.80 \dfrac{m}{s^2} \cdot 6.00 \text{ m} = 85,300$ J

$E_{K6} = ?$

$E_{G16} + E_{K16} = E_{G6} + E_{K6}$

$E_{K6} = E_{G16} + E_{K16} - E_{G6} = 227,360 \text{ J} + 0 - 85,260$ J

$E_{K6} = 142,100$ J

$v_6 = ?$

$v_6 = \sqrt{\dfrac{2E_{K6}}{m}} = \sqrt{\dfrac{2 \cdot 142,100 \text{ J}}{1450 \text{ kg}}} = 14.0 \dfrac{m}{s}$

4.a. 7th calc

$m = 1450$ kg

$g = 9.80 \dfrac{m}{s^2}$

$h = 4.0$ m

$E_{G4} = ?$

$E_{G4} = mgh = 1450 \text{ kg} \cdot 9.80 \dfrac{m}{s^2} \cdot 4.00 \text{ m} = 56,800$ J

$E_{K4} = ?$

$E_{G16} + E_{K16} = E_{G4} + E_{K4}$

$E_{K4} = E_{G16} + E_{K16} - E_{G4} = 227,360 \text{ J} + 0 - 56,840$ J

$E_{K4} = 170,520$ J

$v_4 = ?$

$v_4 = \sqrt{\dfrac{2E_{K4}}{m}} = \sqrt{\dfrac{2 \cdot 170,520 \text{ J}}{1450 \text{ kg}}} = 15.3 \dfrac{m}{s}$

4.a. 8th calc

$m = 1450$ kg

$g = 9.80 \ \dfrac{\text{m}}{\text{s}^2}$

$h = 2.0$ m

$E_{G2} = ?$

$E_{G2} = mgh = 1450 \text{ kg} \cdot 9.80 \ \dfrac{\text{m}}{\text{s}^2} \cdot 2.00 \text{ m} = 28{,}420 \text{ J}$

$E_{K2} = ?$

$E_{G16} + E_{K16} = E_{G2} + E_{K2}$

$E_{K2} = E_{G16} + E_{K16} - E_{G2} = 227{,}360 \text{ J} + 0 - 28{,}420 \text{ J}$

$E_{K2} = 198{,}940 \text{ J}$

$v_2 = ?$

$v_2 = \sqrt{\dfrac{2E_{K2}}{m}} = \sqrt{\dfrac{2 \cdot 198{,}940 \text{ J}}{1450 \text{ kg}}} = 16.6 \ \dfrac{\text{m}}{\text{s}}$

4.a. 9th calc

$m = 1450$ kg

$g = 9.80 \ \dfrac{\text{m}}{\text{s}^2}$

$h = 0$ m

$E_{G0(at \ h \ = \ 0m)} = 0$

$E_{K0(at \ h \ = \ 0m)} = ?$

$E_{G16} + E_{K16} = E_{G0} + E_{K0}$

$E_{K0} = E_{G16} = 227{,}360 \text{ J}$

$v_0 = ?$

$v_0 = \sqrt{\dfrac{2E_{K0}}{m}} = \sqrt{\dfrac{2 \cdot 227{,}360 \text{ J}}{1450 \text{ kg}}} = 17.7 \ \dfrac{\text{m}}{\text{s}}$

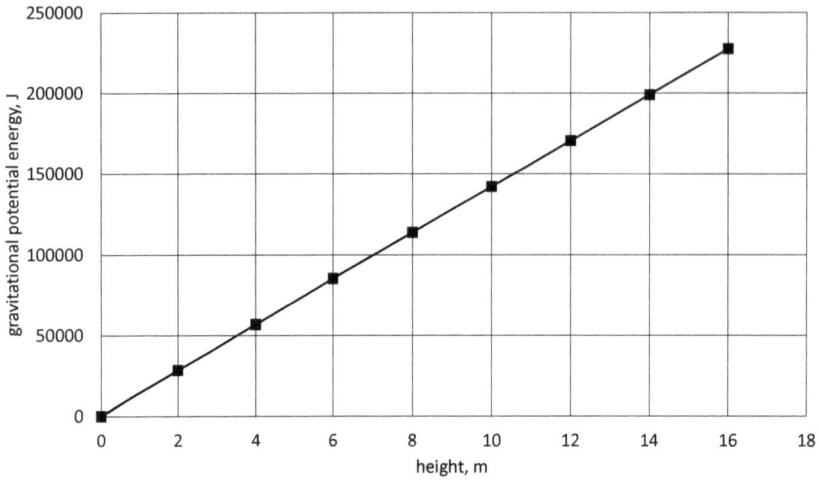

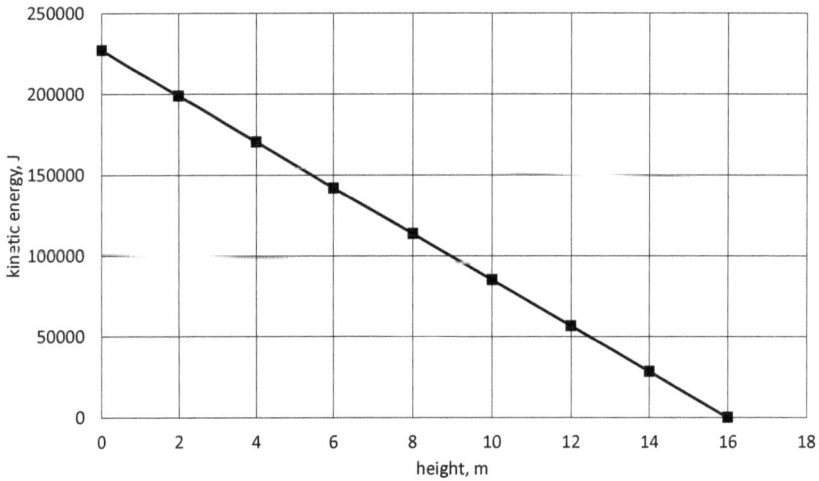

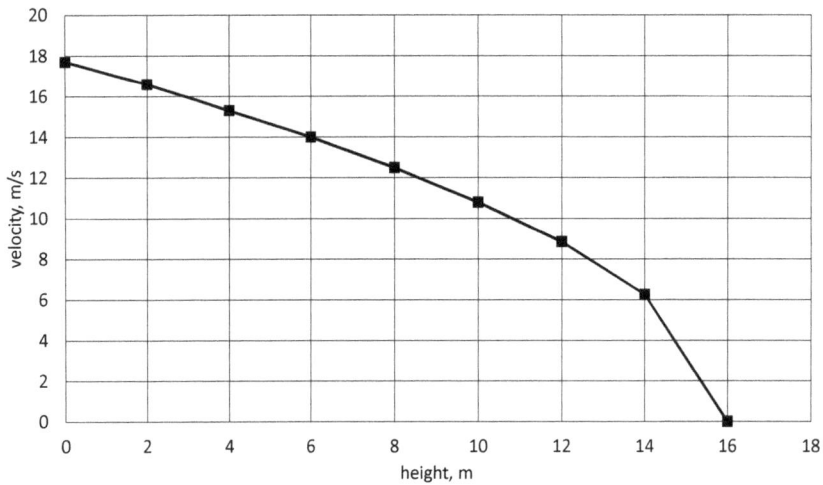

Energy Calculations Set 3

1.a.

$$F_w = 27.05 \text{ lb} \cdot \frac{4.45 \text{ N}}{1 \text{ lb}} = 120.37 \text{ N}$$

$$d = 185 \text{ cm} \cdot \frac{1 \text{ m}}{100 \text{ cm}} = 1.85 \text{ m}$$

$$W = Fd = 120.37 \text{ N} \cdot 1.85 \text{ m} = 223 \text{ J}$$

1.b

$$F_w = 120.37 \text{ N}$$

$$h = 1.85 \text{ m}$$

$$g = 9.80 \frac{\text{m}}{\text{s}^2}$$

$$m = ?$$

$$F_w = mg$$

$$m = \frac{F_w}{g}$$

$$m = \frac{120.37 \text{ N}}{9.80 \frac{\text{m}}{\text{s}^2}} = 12.28 \text{ kg}$$

$$E_G = ?$$

$$E_G = mgh = 12.28 \text{ kg} \cdot 9.80 \frac{\text{m}}{\text{s}^2} \cdot 1.85 \text{ m} = 223 \text{ J}$$

1.c.

$$m = 12.28 \text{ kg}$$

$$g = 9.80 \frac{\text{m}}{\text{s}^2}$$

$$h = 0 \text{ m}$$

$$E_G = ?$$

$$E_G = mgh = 12.28 \text{ kg} \cdot 9.80 \frac{\text{m}}{\text{s}^2} \cdot 0 \text{ m} = 0 \text{ J}$$

1.d.

$E_{Gi} = 222.7 \text{ J}$

$E_{Gf} = 0 \text{ J}$

$E_{Ki} = 0 \text{ J}$

$E_{Kf} = ?$

$E_{Gi} + E_{Ki} = E_{Gf} + E_{Kf}$

$E_{Kf} = E_{Gi} + E_{Ki} - E_{Gf} = 222.7 \text{ J} + 0 - 0 = 223 \text{ J}$

1.e.

$E_{Kf} = 222.7 \text{ J}$

$m = 12.28 \text{ kg}$

$v_f = ?$

$$v_f = \sqrt{\frac{2E_{Kf}}{m}} = \sqrt{\frac{2 \cdot 222.7 \text{ J}}{12.28 \text{ kg}}} = 6.02 \ \frac{\text{m}}{\text{s}}$$

1.f.

$m = 12.28 \text{ kg}$

$g = 9.80 \ \dfrac{\text{m}}{\text{s}^2}$

$h = \dfrac{1}{2} \cdot 1.85 \text{ m} = 0.925 \text{ m}$

$E_{G50(50 \text{ percent down})} = mgh = 12.28 \text{ kg} \cdot 9.80 \ \dfrac{\text{m}}{\text{s}^2} \cdot 0.925 \text{ m} = 111 \text{ J}$

$E_{Kf} = 222.6 \text{ J}$

$E_{G50} + E_{K50} = E_{Gf} + E_{Kf}$

$E_{K50} = E_{Gf} + E_{Kf} - E_{G50} = 0 + 222.6 \text{ J} - 111.3 \text{ J} = 111 \text{ J}$

1.g.

$m = 12.28 \text{ kg}$

$E_{K50} = 111.3 \text{ J}$

$v_{50} = ?$

$$v_{50} = \sqrt{\frac{2E_{K50}}{m}} = \sqrt{\frac{2 \cdot 111.3 \text{ J}}{12.28 \text{ kg}}} = 4.26 \ \frac{\text{m}}{\text{s}}$$

1.h.

$m = 12.28 \text{ kg}$

$g = 9.80 \dfrac{\text{m}}{\text{s}^2}$

$h = 0.10 \cdot 1.85 \text{ m} = 0.185 \text{ m}$

$E_{G90(90 \text{ percent down})} = mgh = 12.28 \text{ kg} \cdot 9.80 \dfrac{\text{m}}{\text{s}^2} \cdot 0.185 \text{ m} = 22.26 \text{ J}$

$E_{Kf} = 222.6 \text{ J}$

$E_{G90} + E_{K90} = E_{Gf} + E_{Kf}$

$E_{K90} = E_{Gf} + E_{Kf} - E_{G90} = 0 + 222.6 \text{ J} - 22.26 = 200.3 \text{ J}$

$v_{90} = ?$

$v_{90} = \sqrt{\dfrac{2E_{K90}}{m}} = \sqrt{\dfrac{2 \cdot 200.3 \text{ J}}{12.28}} = 5.71 \dfrac{\text{m}}{\text{s}}$

2.

$d = 197 \text{ ft} \cdot \dfrac{0.3048 \text{ m}}{1 \text{ ft}} = 60.05 \text{ m}$

$m = 6.016 \times 10^6 \text{ kg}$

$F_w = mg = 6.016 \times 10^6 \text{ kg} \cdot 9.80 \dfrac{\text{m}}{\text{s}^2} = 5.896 \times 10^7 \text{ N}$

$W = ?$

$W = F_w d = 5.896 \times 10^7 \text{ N} \cdot 60.05 \text{ m}$

$W = 3.54 \times 10^9 \text{ J} \cdot \dfrac{1 \text{ GJ}}{1 \times 10^9 \text{ J}} = 3.54 \text{ GJ}$

3.a.

$m = 5122$ kg

$h_A = 25.0$ m

$h_B = 2.5$ m

$h_C = 18.0$ m

$v_B = ?$

$$E_{GA} = mgh_A = 5,122 \text{ kg} \cdot 9.80 \, \frac{m}{s^2} \cdot 25.0 \text{ m} = 1.2549 \times 10^6 \text{ J}$$

$$E_{GB} = mgh_B = 5,122 \text{ kg} \cdot 9.80 \, \frac{m}{s^2} \cdot 2.5 \text{ m} = 1.2549 \times 10^5 \text{ J}$$

$$E_{KA} = 0 \text{ J}$$

$$E_{GA} + E_{KA} = E_{GB} + E_{KB}$$

$$E_{KB} = E_{GA} + E_{KA} - E_{GB} = 1.2549 \times 10^6 \text{ J} + 0 \text{ J} - 1.2549 \times 10^5 \text{ J}$$

$$E_{KB} = 1.1294 \times 10^6 \text{ J}$$

$$v_B = \sqrt{\frac{2E_{KB}}{m}} = \sqrt{\frac{2 \cdot 1.1294 \times 10^6 \text{ J}}{5122 \text{ kg}}} = 21.0 \, \frac{m}{s}$$

3.b.

$m = 5122$ kg

$h_A = 25.0$ m

$h_C = 18.0$ m

$v_C = ?$

$$E_{GA} = 1.2549 \times 10^6 \text{ J}$$

$$E_{GC} = mgh_C = 5,122 \text{ kg} \cdot 9.80 \, \frac{m}{s^2} \cdot 18.0 \text{ m} = 9.0352 \times 10^5 \text{ J}$$

$$E_{KA} = 0 \text{ J}$$

$$E_{GA} + E_{KA} = E_{GC} + E_{KC}$$

$$E_{KC} = E_{GA} + E_{KA} - E_{GC} = 1.2549 \times 10^6 \text{ J} + 0 \text{ J} - 9.0352 \times 10^5 \text{ J}$$

$$E_{KC} = 3.5138 \times 10^5 \text{ J}$$

$$v_C = \sqrt{\frac{2E_{KC}}{m}} = \sqrt{\frac{2 \cdot 3.5138 \times 10^5 \text{ J}}{5122 \text{ kg}}} = 11.7 \, \frac{m}{s}$$

4.a.

$$F_w = 104.6 \text{ lb} \cdot \frac{4.45 \text{ N}}{1 \text{ lb}} = 4.6547 \times 10^2 \text{ N}$$

$$d = 13 \text{ steps} \cdot \frac{16.5 \text{ cm}}{1 \text{ step}} \cdot \frac{1 \text{ m}}{100 \text{ cm}} = 2.145 \text{ m}$$

$$W = ?$$

$$W = F_w d = 4.6547 \times 10^2 \text{ N} \cdot 2.145 \text{ m} = 998 \text{ J}$$

4.b.

$$F_w = 4.6547 \times 10^2 \text{ N}$$

$$F_w = mg$$

$$m = \frac{F_w}{g} = \frac{4.6547 \times 10^2 \text{ N}}{9.80 \frac{\text{m}}{\text{s}^2}} = 47.497 \text{ kg}$$

$$h_i = 2.145 \text{ m}$$

$$v_f = ?$$

$$E_{Gi} = mgh_i = 47.497 \text{ kg} \cdot 9.80 \frac{\text{m}}{\text{s}^2} \cdot 2.145 \text{ m}$$

$$E_{Gi} = 998.43 \text{ J}$$

$$E_{Gi} + E_{Ki} = E_{Gf} + E_{Kf}$$

$$E_{Kf} = E_{Gi} + E_{Ki} - E_{Gf} = 998.43 \text{ J} + 0 - 0$$

$$E_{Kf} = 998.43 \text{ J}$$

$$v_f = \sqrt{\frac{2 E_{Kf}}{m}} = \sqrt{\frac{2 \cdot 998.43 \text{ J}}{47.497 \text{ kg}}} = 6.48 \frac{\text{m}}{\text{s}}$$

5.

$$m = 351 \text{ g} \cdot \frac{1 \text{ kg}}{1000 \text{ g}} = 0.351 \text{ kg}$$

$$v_i = 500.00 \frac{\text{cm}}{\text{s}} \cdot \frac{1 \text{ m}}{100 \text{ cm}} = 5.00 \frac{\text{m}}{\text{s}}$$

$$v_f = 0 \frac{\text{m}}{\text{s}}$$

$$h_i = 0 \text{ m}$$

$$h_f = ?$$

$$E_{Ki} = \tfrac{1}{2}mv^2 = 0.5 \cdot 0.351 \text{ kg} \cdot \left(5.00 \frac{\text{m}}{\text{s}} \right)^2$$

$$E_{Ki} = 4.388 \text{ J}$$

$$E_{Gi} + E_{Ki} = E_{Gf} + E_{Kf}$$

$$E_{Gf} = E_{Gi} + E_{Ki} - E_{Kf} = 0 \text{ J} + 4.388 \text{ J} - 0 \text{ J}$$

$$E_{Gf} = 4.388 \text{ J}$$

$$E_{Gf} = mgh_f$$

$$h_f = \frac{F_{Gf}}{mg} = \frac{4.388 \text{ J}}{0.351 \text{ kg} \cdot 9.80 \frac{\text{m}}{\text{s}^2}} = 1.28 \text{ m}$$

6.

$$F_w = 4294 \text{ lb} \cdot \frac{4.45 \text{ N}}{1 \text{ lb}} = 1.9108 \times 10^4 \text{ N}$$

$$F_w = mg$$

$$m = \frac{F_w}{g} = \frac{1.9108 \times 10^4 \text{ N}}{9.80 \frac{\text{m}}{\text{s}^2}} = 1.9498 \times 10^3 \text{ kg}$$

$$v_i = 27.89 \frac{\text{ft}}{\text{s}} \cdot \frac{0.3048 \text{ m}}{1 \text{ ft}} = 8.5009 \frac{\text{m}}{\text{s}}$$

$$h_f = 7.710 \text{ ft} \cdot \frac{0.3048 \text{ m}}{1 \text{ ft}} = 2.3500 \text{ m}$$

$$v_f = ?$$

$$E_{Ki} = \tfrac{1}{2}mv_i^2 = 0.5 \cdot 1.9498 \times 10^3 \text{ kg} \cdot \left(8.5009 \frac{\text{m}}{\text{s}}\right)^2 = 7.0451 \times 10^4 \text{ J}$$

$$E_{Gf} = mgh = 1.9498 \times 10^3 \text{ kg} \cdot 9.80 \frac{\text{m}}{\text{s}^2} \cdot 2.3500 \text{ m}$$

$$E_{Gf} = 4.4904 \times 10^4 \text{ J}$$

$$E_{Gi} + E_{Ki} = E_{Gf} + E_{Kf}$$

$$E_{Kf} = E_{Gi} + E_{Ki} - E_{Gf} = 0 \text{ J} + 7.0451 \times 10^4 \text{ J} - 4.4904 \times 10^4 \text{ J}$$

$$E_{Kf} = 2.5547 \times 10^4 \text{ J}$$

$$v_f = \sqrt{\frac{2E_{Kf}}{m}} = \sqrt{\frac{2 \cdot 2.5547 \times 10^4 \text{ J}}{1.9498 \times 10^3 \text{ kg}}} = 5.12 \frac{\text{m}}{\text{s}}$$

7.

$h_f = 6.500$ m

$m = 950.0$ g $\cdot \dfrac{1\ \text{kg}}{1000\ \text{g}} = 0.9500$ kg

$v_f = 1.000\ \dfrac{\text{m}}{\text{s}}$

$v_i = ?$

$E_{Gi} = 0$ J

$E_{Gf} = mgh = 0.9500\ \text{kg} \cdot 9.803\ \dfrac{\text{m}}{\text{s}^2} \cdot 6.500\ \text{m} = 60.534$ J

$E_{Kf} = \frac{1}{2}mv_f^{\,2} = 0.5 \cdot 0.9500\ \text{kg} \cdot \left(1.000\ \dfrac{\text{m}}{\text{s}}\right)^2 = 0.4750$ J

$E_{Gi} + E_{Ki} = E_{Gf} + E_{Kf}$

$E_{Ki} = E_{Gf} + E_{Kf} - E_{Gi} = 60.534\ \text{J} + 0.4750\ \text{J} - 0\ \text{J}$

$E_{Ki} = 61.009$ J

$v_i = \sqrt{\dfrac{2E_{Ki}}{m}} = \sqrt{\dfrac{2 \cdot 61.009\ \text{J}}{0.9500\ \text{kg}}} = 11.33\ \dfrac{\text{m}}{\text{s}}$

Energy Calculations Set 4

1.

$$F_w = 3420.1 \text{ lb} \cdot \frac{4.45 \text{ N}}{1 \text{ lb}} = 1.52194 \times 10^4 \text{ N}$$

$$h_i = 31.0 \text{ ft} \cdot \frac{0.3048 \text{ m}}{1 \text{ ft}} = 9.449 \text{ m}$$

$$h_f = 6.50 \text{ ft} \cdot \frac{0.3048 \text{ m}}{1 \text{ ft}} = 1.981 \text{ m}$$

$$v_f = ?$$

$$F_w = mg$$

$$m = \frac{F_w}{g} = \frac{1.52194 \times 10^4 \text{ N}}{9.80 \ \frac{\text{m}}{\text{s}^2}} = 1.5530 \times 10^3 \text{ kg}$$

$$E_{Gi} = mgh_i = 1.5530 \times 10^3 \text{ kg} \cdot 9.80 \ \frac{\text{m}}{\text{s}^2} \cdot 9.449 \text{ m} = 1.4381 \times 10^5 \text{ J}$$

$$E_{Gf} = mgh_f = 1.5530 \times 10^3 \text{ kg} \cdot 9.80 \ \frac{\text{m}}{\text{s}^2} \cdot 1.981 \text{ m} = 3.0150 \times 10^4 \text{ J}$$

$$E_{Ki} = 0 \text{ J}$$

$$E_{Gi} + E_{Ki} = E_{Gf} + E_{Kf}$$

$$E_{Kf} = E_{Gi} + E_{Ki} - E_{Gf} = 1.4381 \times 10^5 \text{ J} + 0 \text{ J} - 3.0150 \times 10^4 \text{ J}$$

$$E_{Kf} = 1.1366 \times 10^5 \text{ J}$$

$$v_f = \sqrt{\frac{2E_{Kf}}{m}} = \sqrt{\frac{2 \cdot 1.1366 \times 10^5 \text{ J}}{1.5530 \times 10^3 \text{ kg}}} = 12.1 \ \frac{\text{m}}{\text{s}}$$

2.

$m = 2200$ kg

$h_f = 33.87$ m

$h_i = 37.00$ m

$v_f = 17.88 \dfrac{m}{s}$

$v_i = ?$

$E_{Gi} = mgh_i = 2200 \text{ kg} \cdot 9.80 \ \dfrac{m}{s^2} \cdot 37.00 \text{ m} = 7.98 \times 10^5 \text{ J}$

$E_{Gf} = mgh_f = 2200 \text{ kg} \cdot 9.80 \ \dfrac{m}{s^2} \cdot 33.87 \text{ m} = 7.31 \times 10^5 \text{ J}$

$E_{Kf} = \tfrac{1}{2}mv^2 = 0.5 \cdot 2200 \text{ kg} \cdot (17.88 \ \dfrac{m}{s})^2 = 3.52 \times 10^5 \text{ J}$

$E_{Gi} + E_{Ki} = E_{Gf} + E_{Kf}$

$E_{Ki} = E_{Gf} + E_{Kf} - E_{Gi}$

$E_{Ki} = 7.31 \times 10^5 \text{ J} + 3.52 \times 10^5 \text{ J} - 7.98 \times 10^5 \text{ J} = 2.85 \times 10^5 \text{ J}$

$v_i - \sqrt{\dfrac{2E_{Ki}}{m}} = \sqrt{\dfrac{2 \cdot 2.85 \times 10^5 \text{ J}}{2200 \text{ kg}}} = 16 \ \dfrac{m}{s}$

3.

$$m = 1.873 \times 10^{-2} \text{ mg} \cdot \frac{1 \text{ g}}{1000 \text{ mg}} \cdot \frac{1 \text{ kg}}{1000 \text{ g}} = 1.873 \times 10^{-8} \text{ kg}$$

$$h_i = 2.177 \text{ cm} \cdot \frac{1 \text{ m}}{100 \text{ cm}} = 0.02177 \text{ m}$$

$$v_i = 202.75 \frac{\text{cm}}{\text{s}} \cdot \frac{1 \text{ m}}{100 \text{ cm}} = 2.0275 \frac{\text{m}}{\text{s}}$$

$$h_f = 7.50 \text{ cm} \cdot \frac{1 \text{ m}}{100 \text{ cm}} = 0.0750 \text{ m}$$

$$v_f = ?$$

$$E_{Gi} = mgh_i = 1.873 \times 10^{-8} \text{ kg} \cdot 9.80 \frac{\text{m}}{\text{s}^2} \cdot 0.02177 \text{ m} = 3.996 \times 10^{-9} \text{ J}$$

$$E_{Ki} = \tfrac{1}{2}mv^2 = 0.5 \cdot 1.873 \times 10^{-8} \text{ kg} \cdot \left(2.0275 \frac{\text{m}}{\text{s}}\right)^2 = 3.850 \times 10^{-8} \text{ J}$$

$$E_{Gf} = mgh_f = 1.873 \times 10^{-8} \text{ kg} \cdot 9.80 \frac{\text{m}}{\text{s}^2} \cdot 0.0750 \text{ m} = 1.377 \times 10^{-8} \text{ J}$$

$$E_{Gi} + E_{Ki} = E_{Gf} + E_{Kf}$$

$$E_{Kf} = E_{Gi} + E_{Ki} - E_{Gf} = 3.996 \times 10^{-9} \text{ J} + 3.850 \times 10^{-8} \text{ J} - 1.377 \times 10^{-8} \text{ J}$$

$$E_{Kf} = 2.873 \times 10^{-8} \text{ J}$$

$$v_f = \sqrt{\frac{2E_{Kf}}{m}} = \sqrt{\frac{2 \cdot 2.873 \times 10^{-8} \text{ J}}{0.0750 \text{ m}}} = 1.75 \frac{\text{m}}{\text{s}}$$

4.

$$m = 9.0022 \ \mu g \cdot \frac{1 \ g}{10^6 \ \mu g} \cdot \frac{1 \ kg}{1000 \ g} = 9.0022 \times 10^{-9} \ kg$$

$$h_i = 9.0125 \ cm \cdot \frac{1 \ m}{100 \ cm} = 0.090125 \ m$$

$$v_i = 0$$

$$v_f = 85.160 \ \frac{cm}{s} \cdot \frac{1 \ m}{100 \ cm} = 0.85160 \ \frac{m}{s}$$

$$h_f = ?$$

$$E_{Ki} + E_{Gi} = E_{Kf} + E_{Gf}$$

$$E_{Gf} = E_{Ki} + E_{Gi} - E_{Kf}$$

$$E_{Ki} = 0$$

$$E_{Gi} = mgh_i = 9.0022 \times 10^{-9} \ kg \cdot 9.80 \ \frac{m}{s^2} \cdot 0.090125 \ m = 7.951 \times 10^{-9} \ J$$

$$E_{Kf} = \tfrac{1}{2}mv_f^2 = 0.5 \cdot 9.0022 \times 10^{-9} \ kg \cdot \left(0.85160 \ \frac{m}{s}\right)^2 = 3.264 \times 10^{-9} \ J$$

$$E_{Gf} = 0 + 7.951 \times 10^{-9} \ J - 3.264 \times 10^{-9} \ J = 4.687 \times 10^{-9} \ J$$

$$E_{Gf} = mgh_f$$

$$h_f = \frac{E_{Gf}}{mg} = \frac{4.687 \times 10^{-9} \ J}{9.0022 \times 10^{-9} \ kg \cdot 9.80 \ \frac{m}{s^2}} = 0.0531 \ m \cdot \frac{100 \ cm}{1 \ m} = 5.31 \ cm$$

5.

$$v_i = 1.562 \; \frac{\text{ft}}{\text{s}} \cdot \frac{0.3048 \text{ m}}{1 \text{ ft}} = 0.40761 \; \frac{\text{m}}{\text{s}}$$

$$F_w = 4.1843 \text{ lb} \cdot \frac{4.45 \text{ N}}{1 \text{ lb}} = 18.6201 \text{ N}$$

$$v_f = 24.75 \; \frac{\text{ft}}{\text{s}} \cdot \frac{0.3048 \text{ m}}{1 \text{ ft}} = 7.5438 \; \frac{\text{m}}{\text{s}}$$

$$h_f = 0 \text{ m}$$

$$h_i = ?$$

$$F_w = mg$$

$$m = \frac{F_w}{g} = \frac{18.6201 \text{ N}}{9.80 \; \frac{\text{m}}{\text{s}^2}} = 1.900 \text{ kg}$$

$$E_{Ki} = \tfrac{1}{2}mv_i^2 = 0.5 \cdot 1.900 \text{ kg} \cdot \left(0.4761 \; \frac{\text{m}}{\text{s}}\right)^2 = 0.2153 \text{ J}$$

$$E_{Gf} = 0 \text{ J}$$

$$E_{Kf} = \tfrac{1}{2}mv_f^2 = 0.5 \cdot 1.900 \text{ kg} \cdot \left(7.5438 \; \frac{\text{m}}{\text{s}}\right)^2 = 54.063 \text{ J}$$

$$E_{Gi} + E_{Ki} = E_{Gf} + E_{Kf}$$

$$E_{Gi} = E_{Gf} + E_{Kf} - E_{Ki} = 0 \text{ J} + 54.063 \text{ J} - 0.2153 \text{ J}$$

$$E_{Gi} = 53.85 \text{ J}$$

$$E_{Gi} = mgh_i$$

$$h_i = \frac{E_{Gi}}{mg} = \frac{53.85 \text{ J}}{1.900 \text{ kg} \cdot 9.80 \; \frac{\text{m}}{\text{s}^2}} = 2.89 \text{ m}$$

Chapter 6

Temperature Unit Conversions

1.a.

$T_C = 32.0°C$

$T_F = ?$

$T_C = \dfrac{5}{9}(T_F - 32°)$

$T_F = \dfrac{9}{5}T_C + 32° = \dfrac{9}{5} \cdot 32.0°C + 32° = 89.6°F$

1.b.

$T_C = 32.0°C$

$T_K = ?$

$T_K = T_C + 273.2$

$T_K = 32.0°C + 273.2 = 305.2 \text{ K}$

2.a.

$T_F = 56.5°F$

$T_C = ?$

$T_C = \dfrac{5}{9}(T_F - 32°)$

$T_C = \dfrac{5}{9}(56.5°F - 32°) = 13.6°C$

2.b.

$T_F = 56.5°F$

$T_C = 13.6°C$

$T_K = ?$

$T_K = T_C + 273.2$

$T_K = 13.6°C + 273.2 = 286.8 \text{ K}$

3.a.

$T_K = 455.0 \text{ K}$

$T_C = 181.8°C$

$T_F = ?$

$T_C = \dfrac{5}{9}(T_F - 32°)$

$T_F = \dfrac{9}{5}T_C + 32°$

$T_F = \dfrac{9}{5} \cdot 181.8°C + 32° = 359.2°F$

3.b.

$T_K = 455.0 \text{ K}$

$T_C = ?$

$T_K = T_C + 273.2$

$T_C = T_K - 273.2 = 455.0 \text{ K} - 273.2 = 181.8°C$

4.a.

$T_F = -17.9°F$

$T_C = ?$

$T_C = \dfrac{5}{9}(T_F - 32°)$

$T_C = \dfrac{5}{9}(-17.9°F - 32°) = -27.7°C$

4.b.

$T_F = -17.9°F$

$T_C = -27.7°C$

$T_K = ?$

$T_K = T_C + 273.2$

$T_K = -27.7°C + 273.2 = 245.5 \text{ K}$

5.a.

$T_C = -41.6°C$

$T_F = ?$

$$T_C = \frac{5}{9}(T_F - 32°)$$

$$T_F = \frac{9}{5}T_C + 32°$$

$$T_F = \left(\frac{9}{5}\cdot(-41.6°C)\right) + 32° = -42.9°F$$

5.b.

$T_C = -41.6°C$

$T_K = ?$

$T_K = T_C + 273.2$

$T_K = -41.6°C + 273.2 = 231.6 \text{ K}$

6.a.

$T_K = 79.0 \text{ K}$

$T_C = -194.2°C$

$T_F = ?$

$$T_C = \frac{5}{9}(T_F - 32°)$$

$$T_F = \frac{9}{5}T_C + 32°$$

$$T_F = \left(\frac{9}{5}\cdot(-194.2°C)\right) + 32° = -317.6° \text{ F}$$

6.b.

$T_K = 79.0 \text{ K}$

$T_C = ?$

$T_K = T_C + 273.2$

$T_C = T_K - 273.2$

$T_C = 79.0 \text{ K} - 273.2 = -194.2°C$

Chapter 7

1.

$f = 60.0$ Hz

$\tau = ?$

$\tau = \dfrac{1}{f}$

$\tau = \dfrac{1}{60.0 \text{ Hz}} = 0.0167$ s

2.

$\tau = 2.155 \times 10^{-5}$ s

$f = ?$

$\tau = \dfrac{1}{f}$

$f = \dfrac{1}{\tau} = \dfrac{1}{2.155 \times 10^{-5} \text{ s}} = 46{,}404 \text{ Hz} \cdot \dfrac{1 \text{ kHz}}{1000 \text{ Hz}} = 46.40$ kHz

---

3.

$f = 26.0 \text{ kHz} \cdot \dfrac{1000 \text{ Hz}}{1 \text{ kHz}} = 2.60 \times 10^{4}$ Hz

$\tau = ?$

$\tau = \dfrac{1}{f} = \dfrac{1}{2.60 \times 10^{4} \text{ Hz}} = 3.85 \times 10^{-5}$ s

---

4.

$f = 2.60 \times 10^{4}$ Hz

$v = 342 \dfrac{\text{m}}{\text{s}}$

$\lambda = ?$

$v = \lambda f$

$\lambda = \dfrac{v}{f} = \dfrac{342 \dfrac{\text{m}}{\text{s}}}{2.60 \times 10^{4} \text{ Hz}} = 1.32 \times 10^{-2} \text{ m} \cdot \dfrac{100 \text{ cm}}{1 \text{ m}} = 1.32 \text{ cm}$

5.

$$f = 89.5 \text{ MHz} \cdot \frac{1 \times 10^6 \text{ Hz}}{1 \text{ MHz}} = 8.95 \times 10^7 \text{ Hz}$$

$$v = c = 3.00 \times 10^8 \ \frac{\text{m}}{\text{s}}$$

$$\lambda = ?$$

$$\tau = ?$$

$$v = \lambda f$$

$$\lambda = \frac{v}{f} = \frac{3.00 \times 10^8 \ \frac{\text{m}}{\text{s}}}{8.95 \times 10^7 \text{ Hz}} = 3.35 \text{ m}$$

$$\tau = \frac{1}{f} = \frac{1}{8.95 \times 10^7 \text{ Hz}} = 1.12 \times 10^{-8} \text{ s} \cdot \frac{1 \times 10^6 \ \mu\text{s}}{1 \text{ s}} = 0.0112 \ \mu\text{s}$$

6.

$$v = c = 3.00 \times 10^8 \ \frac{\text{m}}{\text{s}}$$

$$f = 1310 \text{ kHz} \cdot \frac{1000 \text{ Hz}}{1 \text{ kHz}} = 1.31 \times 10^6 \text{ Hz}$$

$$\lambda = ?$$

$$\tau = ?$$

$$v = \lambda f$$

$$\lambda = \frac{v}{f} = \frac{3.00 \times 10^8 \ \frac{\text{m}}{\text{s}}}{1.31 \times 10^6 \text{ Hz}} = 229 \text{ m}$$

$$\tau = \frac{1}{f} = \frac{1}{1.31 \times 10^6 \text{ Hz}} = 7.64 \times 10^{-7} \text{ s} \cdot \frac{1 \times 10^6 \ \mu\text{s}}{1 \text{ s}} = 0.763 \ \mu\text{s}$$

7.

$$v = c = 3.00 \times 10^8 \ \frac{m}{s}$$

$$\lambda = 542 \ \text{nm} \cdot \frac{1 \ m}{1 \times 10^9 \ \text{nm}} = 5.42 \times 10^{-7} \ m$$

$$\tau = ?$$

$$f = ?$$

$$v = \lambda f$$

$$f = \frac{v}{\lambda} = \frac{3.00 \times 10^8 \ \frac{m}{s}}{5.42 \times 10^{-7} \ m} = 5.54 \times 10^{14} \ \text{Hz} \cdot \frac{1 \ \text{THz}}{1 \times 10^{12} \ \text{Hz}} = 554 \ \text{THz}$$

$$\tau = \frac{1}{f} = \frac{1}{5.54 \times 10^{14} \ \text{Hz}} = 1.81 \times 10^{-15} \ \text{s} \cdot \frac{1 \times 10^{12} \ \text{ps}}{1 \ \text{s}} = 0.00181 \ \text{ps}$$

8.

$$\lambda = 10.6 \ \mu\text{m} \cdot \frac{1 \ m}{1 \times 10^6 \ \mu\text{m}} = 1.06 \times 10^{-5} \ m$$

$$v = 3.00 \times 10^8 \ \frac{m}{s}$$

$$\tau = ?$$

$$f = ?$$

$$v = \lambda f$$

$$f = \frac{v}{\lambda} = \frac{3.00 \times 10^8 \ \frac{m}{s}}{1.06 \times 10^{-5} \ m} = 2.83 \times 10^{13} \ \text{Hz} \cdot \frac{1 \ \text{GHz}}{1 \times 10^9 \ \text{Hz}} = 2.83 \times 10^4 \ \text{GHz}$$

$$\tau = \frac{1}{f} = \frac{1}{2.83 \times 10^{13} \ \text{Hz}} = 3.53 \times 10^{-14} \ \text{s} \cdot \frac{1 \times 10^9 \ \text{ns}}{1 \ \text{s}} = 3.53 \times 10^{-5} \ \text{ns}$$

9.

$$f = 33 \text{ kHz} \cdot \frac{1000 \text{ Hz}}{1 \text{ kHz}} = 3.3 \times 10^4 \text{ Hz}$$

$$v = 342 \ \frac{\text{m}}{\text{s}}$$

$$\tau = ?$$

$$\lambda = ?$$

$$\tau = \frac{1}{f} = \frac{1}{3.3 \times 10^4 \text{ Hz}} = 3.03 \times 10^{-5} \text{ s} \cdot \frac{1000 \text{ ms}}{1 \text{ s}} = 0.030 \text{ ms}$$

$$v = \lambda f$$

$$\lambda = \frac{v}{f} = \frac{342 \ \frac{\text{m}}{\text{s}}}{3.3 \times 10^4 \text{ Hz}} = 1.04 \times 10^{-2} \text{ m} \cdot \frac{1000 \text{ mm}}{1 \text{ m}} = 1.0 \times 10^1 \text{ mm}$$

10.

Wavelengths smaller than the hole size pass through the screen.

$$v = 3.00 \times 10^8 \ \frac{\text{m}}{\text{s}}$$

$$\lambda = 2.00 \text{ mm} \cdot \frac{1 \text{ m}}{1000 \text{ mm}} = 2.00 \times 10^{-3} \text{ m}$$

$$v = \lambda f$$

$$f = \frac{v}{\lambda} = \frac{3.00 \times 10^8 \ \frac{\text{m}}{\text{s}}}{2.00 \times 10^{-3} \text{ m}} = 1.50 \times 10^{11} \text{ Hz}$$

Thus, when $f > 1.50 \times 10^{11}$ Hz, the waves pass through.

11.

$$\lambda_b = 470 \text{ nm} \cdot \frac{1 \text{ m}}{1 \times 10^9 \text{ nm}} = 4.7 \times 10^{-7} \text{ m}$$

$$\lambda_g = 550 \text{ nm} \cdot \frac{1 \text{ m}}{1 \times 10^9 \text{ nm}} = 5.5 \times 10^{-7} \text{ m}$$

$$\lambda_r = 680 \text{ nm} \cdot \frac{1 \text{ m}}{1 \times 10^9 \text{ nm}} = 6.8 \times 10^{-7} \text{ m}$$

$$v = 3.00 \times 10^8 \ \frac{\text{m}}{\text{s}}$$

$$f_b = ?$$

$$f_g = ?$$

$$f_r = ?$$

$$v = \lambda f$$

$$f = \frac{v}{\lambda}$$

$$f_b = \frac{3.00 \times 10^8 \ \frac{\text{m}}{\text{s}}}{4.7 \times 10^{-7} \text{ m}} = 6.4 \times 10^{14} \text{ Hz}$$

$$f_g = \frac{3.00 \times 10^8 \ \frac{\text{m}}{\text{s}}}{5.5 \times 10^{-7} \text{ m}} = 5.5 \times 10^{14} \text{ Hz}$$

$$f_r = \frac{3.00 \times 10^8 \ \frac{\text{m}}{\text{s}}}{6.8 \times 10^{-7} \text{ m}} = 4.4 \times 10^{14} \text{ Hz}$$

12.

$$f = 20.00 \text{ Hz}$$

$$v = 342.0 \ \frac{\text{m}}{\text{s}}$$

$$\lambda = ?$$

$$v = \lambda f$$

$$\lambda = \frac{v}{f} = \frac{342.0 \ \frac{\text{m}}{\text{s}}}{20.00 \text{ Hz}} = 17.10 \text{ m}$$

13.a.

$f = 4.67 \times 10^{20}$ Hz

$v = 3.00 \times 10^8 \ \dfrac{m}{s}$

$\lambda = ?$

$v = \lambda f$

$$\lambda = \frac{v}{f} = \frac{3.00 \times 10^8 \ \dfrac{m}{s}}{4.67 \times 10^{20} \ Hz} = 6.42 \times 10^{-13} \ m \cdot \frac{1 \times 10^{12} \ pm}{m} = 0.642 \ pm$$

13.b.

$f = 9.9876 \times 10^{18}$ Hz

$v = 3.00 \times 10^8 \ \dfrac{m}{s}$

$\lambda = ?$

$v = \lambda f$

$$\lambda = \frac{v}{f} = \frac{3.00 \times 10^8 \ \dfrac{m}{s}}{9.9876 \times 10^{18} \ Hz} = 3.00 \times 10^{-11} \ m \cdot \frac{1 \times 10^{9} \ nm}{m} = 0.0300 \ nm$$

13.c.

$f = 2.555 \times 10^{10}$ Hz

$v = 3.00 \times 10^8 \ \dfrac{m}{s}$

$\lambda = ?$

$v = \lambda f$

$$\lambda = \frac{v}{f} = \frac{3.00 \times 10^8 \ \dfrac{m}{s}}{2.555 \times 10^{10} \ Hz} = 0.0117 \ m \cdot \frac{1000 \ mm}{m} = 11.7 \ mm$$

13.d.

$f = 1.172 \times 10^{15}$ Hz

$v = 3.00 \times 10^8 \ \dfrac{\text{m}}{\text{s}}$

$\lambda = ?$

$v = \lambda f$

$$\lambda = \frac{v}{f} = \frac{3.00 \times 10^8 \ \dfrac{\text{m}}{\text{s}}}{1.172 \times 10^{15} \ \text{Hz}} = 2.56 \times 10^{-7} \ \text{m} \cdot \frac{1 \times 10^9 \ \text{nm}}{\text{m}} = 256 \ \text{nm}$$

13.e

$f = 2.83 \times 10^{13}$ Hz

$v = 3.00 \times 10^8 \ \dfrac{\text{m}}{\text{s}}$

$\lambda = ?$

$v = \lambda f$

$$\lambda = \frac{v}{f} = \frac{3.00 \times 10^8 \ \dfrac{\text{m}}{\text{s}}}{2.83 \times 10^{13} \ \text{Hz}} = 1.06 \times 10^{-5} \ \text{m} \cdot \frac{1 \times 10^6 \ \mu\text{m}}{\text{m}} = 10.6 \ \mu\text{m}$$

14.

$$f = 2.45 \ \text{GHz} \cdot \frac{1 \times 10^9 \ \text{Hz}}{1 \ \text{GHz}} = 2.45 \times 10^9 \ \text{Hz}$$

$$D = 3 \ \text{mm} \cdot \frac{1 \ \text{m}}{10^3 \ \text{mm}} = 0.003 \ \text{m}$$

$v = \lambda f$

$$\lambda = \frac{v}{f} = \frac{3.00 \times 10^8 \ \dfrac{\text{m}}{\text{s}}}{2.45 \times 10^9 \ \text{Hz}} = 0.12 \ \text{m}$$

$ratio = ?$

$$ratio = \frac{\lambda}{D} = \frac{0.12 \ \text{m}}{0.003 \ \text{m}} = 40$$

15.a.

$$f = 1.00 \text{ kHz} \cdot \frac{1000 \text{ Hz}}{1 \text{ kHz}} = 1.00 \times 10^3 \text{ Hz}$$

$$v = 342 \ \frac{\text{m}}{\text{s}}$$

$$\lambda = ?$$

$$v = \lambda f$$

$$\lambda = \frac{v}{f} = \frac{342 \ \frac{\text{m}}{\text{s}}}{1.00 \times 10^3 \text{ Hz}} = 0.342 \text{ m}$$

15.b.

$$f = 1.00 \text{ kHz} \cdot \frac{1000 \text{ Hz}}{1 \text{ kHz}} = 1.00 \times 10^3 \text{ Hz}$$

$$v = 1402 \ \frac{\text{m}}{\text{s}}$$

$$\lambda = ?$$

$$v = \lambda f$$

$$\lambda = \frac{v}{f} = \frac{1402 \ \frac{\text{m}}{\text{s}}}{1.00 \times 10^3 \text{ Hz}} = 1.40 \text{ m}$$

15.c.

$$f = 1.00 \text{ kHz} \cdot \frac{1000 \text{ Hz}}{1 \text{ kHz}} = 1.00 \times 10^3 \text{ Hz}$$

$$v = 5130 \ \frac{\text{m}}{\text{s}}$$

$$\lambda = ?$$

$$v = \lambda f$$

$$\lambda = \frac{v}{f} = \frac{5130 \ \frac{\text{m}}{\text{s}}}{1.00 \times 10^3 \text{ Hz}} = 5.13 \text{ m}$$

15.d.

$$f = 1.00 \text{ kHz} \cdot \frac{1000 \text{ Hz}}{1 \text{ kHz}} = 1.00 \times 10^3 \text{ Hz}$$

$$v = 965 \ \frac{\text{m}}{\text{s}}$$

$$\lambda = ?$$

$$v = \lambda f$$

$$\lambda = \frac{v}{f} = \frac{965 \ \dfrac{\text{m}}{\text{s}}}{1.00 \times 10^3 \text{ Hz}} = 0.965 \text{ m}$$

Chapter 8

Introductory Circuit Calculations

1.

$I = 13.00$ A

$V = 25.00$ V

$R = ?$

$V = IR$

$$R = \frac{V}{I} = \frac{25.00 \text{ V}}{13.00 \text{ A}} = 1.923 \ \Omega$$

2.

$V = 24$ V

$R = 250 \ \Omega$

$I = ?$

$V = IR$

$$I = \frac{V}{R} = \frac{24 \text{ V}}{250 \ \Omega} = 0.096 \text{ A} \cdot \frac{1000 \text{ mA}}{1 \text{ A}} = 96 \text{ mA}$$

3.

$R = 12.20$ kΩ

$V = 4.500$ V

$I = ?$

$V = IR$

$$I = \frac{V}{R} = \frac{4.500 \text{ V}}{12.20 \text{ k}\Omega} = 0.3689 \text{ mA}$$

4.

$I = 0.0300$ mA

$$R = 33.3 \text{ M}\Omega \cdot \frac{1 \times 10^3 \text{ k}\Omega}{1 \text{ M}\Omega} = 3.33 \times 10^4 \text{ k}\Omega$$

$V = ?$

$V = IR = 0.0300 \text{ mA} \cdot 3.33 \times 10^4 \text{ k}\Omega = 999$ V

5.a.

$I = 13.00$ A

$V = 25.00$ V

$P = ?$

$P = VI = 25.00$ V $\cdot 13.00$ A $= 325.0$ W

5.b.

$V = 24$ V

$I = 96$ mA

$P = ?$

$P = VI = 24$ V $\cdot 96$ mA $= 2300$ mW $= 2.3$ W

5.c.

$V = 4.500$ V

$I = 0.3689$ mA

$P = ?$

$P = VI = 4.500$ V $\cdot 0.3689$ mA $= 1.660$ mW

5.d.

$I = 0.0300$ mA

$V = 999$ V

$P = ?$

$P = VI = 999$ V $\cdot 0.0300$ mA $= 30.0$ mW

6.

$V = 120$ V

$P = 60.00$ W

$R = ?$

$I = ?$

$P = VI$

$$I = \frac{P}{V} = \frac{60.00 \text{ W}}{120 \text{ V}} = 0.50 \text{ A}$$

$V = IR$

$$R = \frac{V}{I} = \frac{120 \text{ V}}{0.50 \text{ A}} = 240 \ \Omega$$

7.

$V = 120$ V

$I = 12$ A

$P = ?$

$$P = VI = 120 \text{ V} \cdot 12 \text{ A} = 1440 \text{ W} \cdot \frac{1 \text{ kW}}{1000 \text{ W}} = 1.4 \text{ kW}$$

8.

$I = 13.5$ μA

$V = 6.0$ V

$P = ?$

$P = VI = 6.0 \text{ V} \cdot 13.5 \text{ μA} = 81 \text{ μW}$

9.

$$P = 155 \text{ MW} \cdot \frac{1 \times 10^6 \text{ W}}{1 \text{ MW}} = 1.55 \times 10^8 \text{ W}$$

$V = 762$ V

$I = ?$

$P = VI$

$$I = \frac{P}{V} = \frac{1.55 \times 10^8 \text{ W}}{762 \text{ V}} = 203{,}000 \text{ A}$$

Relationships Between Variables in Electric Circuits

voltage, V	current, mA
0	0
2	0.606
4	1.21
6	1.82
8	2.42
10	3.03
12	3.64
14	4.24
15	4.55

Table 1. Current vs. voltage for 3.3 kΩ resistor.

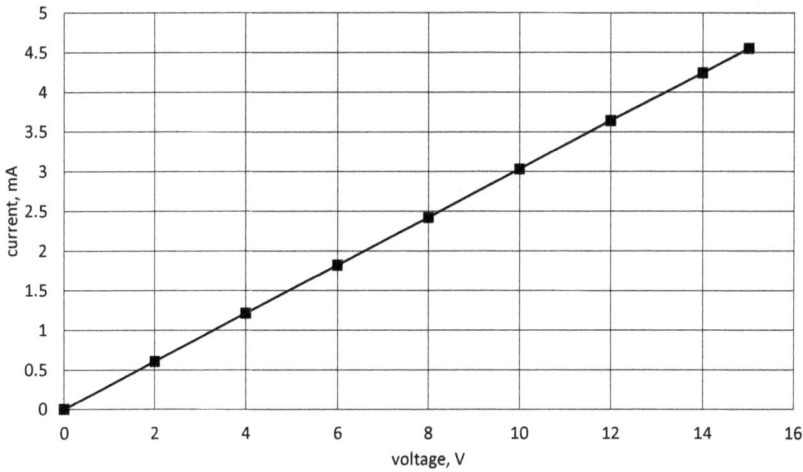

Graph 1. Current vs. voltage for 3.3 kΩ resistor.

current, mA	power, mW
0	0
0.606	1.212
1.21	4.84
1.82	10.92
2.42	19.36
3.03	30.30
3.64	43.68
4.24	59.36
4.55	68.25

Table 2. Power vs. current for 3.3 kΩ resistor.

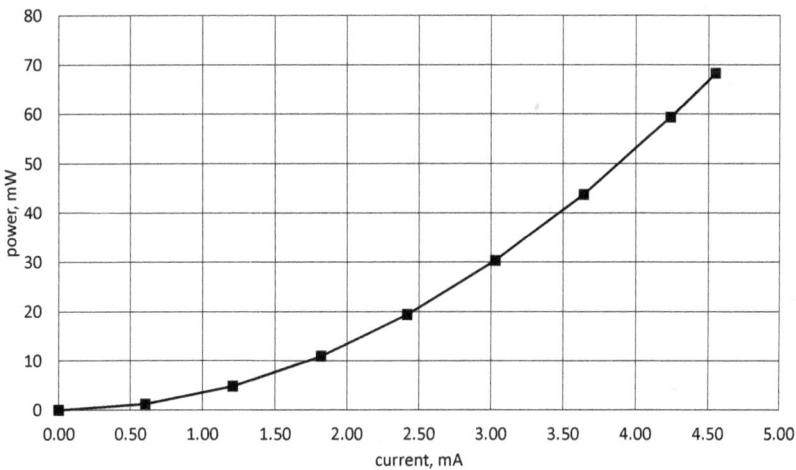

Graph 2. Power vs. current for 3.3 kΩ resistor.

Equivalent Resistance Exercises

1.

series
$$980\ \Omega + 440\ \Omega = 1420\ \Omega$$

2.

series
$$13\ k\Omega + 22\ k\Omega = 35\ k\Omega$$

3.

parallel
$$\frac{5.1\ M\Omega \cdot 0.99\ M\Omega}{5.1\ M\Omega + 0.99\ M\Omega} = 0.8291\ M\Omega$$

4.

parallel
$$\frac{13.5\ k\Omega \cdot 6.2\ k\Omega}{13.5\ k\Omega + 6.2\ k\Omega} = 4.2487\ k\Omega$$

5.

parallel
$$\frac{12\ k\Omega \cdot 12\ k\Omega}{12\ k\Omega + 12\ k\Omega} = 6\ k\Omega$$

Equivalent Resistance Calculations

1.

$$R_3 + R_4 = 47\ \Omega + 130\ \Omega = 177\ \Omega$$
$$R_2 \| 177\ \Omega = \frac{32\ \Omega \cdot 177\ \Omega}{32\ \Omega + 177\ \Omega} = 27.1005\ \Omega$$
$$R_{EQ} = R_1 + 27.1005\ \Omega_4 = 55\ \Omega + 27.1005\ \Omega = 82.1005\ \Omega$$

2.

$$R_3 \| R_4 = \frac{2.2\ k \cdot 0.990\ k}{2.2\ k + 0.990\ k} = 0.6828\ k$$

$$R_2 \| 0.6828\ k = \frac{5.7\ k \cdot 0.6828\ k}{5.7\ k + 0.6828\ k} = 0.6098\ k$$

$$R_{EQ} = R_1 \| 0.6098\ k = \frac{4.1\ k \cdot 0.6098\ k}{4.1\ k + 0.6098\ k} = 0.5308\ k\Omega$$

3.

$$R_3 \| R_4 = \frac{33 \text{ k} \cdot 33 \text{ k}}{33 \text{ k} + 33 \text{ k}} = 16.5 \text{ k}$$

$$R_{EQ} = R_1 + R_2 + 16.5 \text{ k} = 13 \text{ k} + 27 \text{ k} + 16.5 \text{ k} = 56.5000 \text{ k}\Omega$$

4.

$$R_4 + R_5 = 630 \ \Omega + 630 \ \Omega = 1260 \ \Omega$$

$$R_3 \| 1260 \text{ k} = \frac{470 \ \Omega \cdot 1260 \ \Omega}{470 \ \Omega + 1260 \ \Omega} = 342.3121 \ \Omega$$

$$R_2 + 342.3121 \ \Omega = 220 \ \Omega + 342.3121 \ \Omega = 562.3121 \ \Omega$$

$$R_{EQ} = R_1 \| 562.3121 \ \Omega = 540 \ \Omega \| 562.3121 \ \Omega$$

$$R_{EQ} = \frac{540 \ \Omega \cdot 562.3121 \ \Omega}{540 \ \Omega + 562.3121 \ \Omega} = 275.4651 \ \Omega$$

5.

$$R_2 \| R_3 = \frac{0.970 \text{ M} \cdot 0.860 \text{ M}}{0.970 \text{ M} + 0.880 \text{ M}} = 0.4558 \text{ M}$$

$$R_{EQ} = R_1 + 0.4558 \text{ M} + R_4 = 15 \text{ M} + 0.4558 \text{ M} + 11 \text{ M} = 26.4558 \text{ M}\Omega$$

6.

$$R_3 + R_4 = 6.1 \text{ k} + 1.3 \text{ k} = 7.4 \text{ k}$$

$$R_2 \| 7.4 \text{ k} = 2.4 \text{ k} \| 7.4 \text{ k} = \frac{2.4 \text{ k} \cdot 7.4 \text{ k}}{2.4 \text{ k} + 7.4 \text{ k}} = 1.8122 \text{ k}$$

$$R_{EQ} = R_1 \| 1.8122 \text{ k} = 1.5 \text{ k} \| 1.8122 \text{ k}$$

$$R_{EQ} = \frac{1.5 \text{ k} \cdot 1.8122 \text{ k}}{1.5 \text{ k} + 1.8122 \text{ k}} = 0.8207 \text{ k}\Omega$$

7.

$$R_3 + R_4 = 16 \ \Omega + 22 \ \Omega = 38 \ \Omega$$

$$R_2 \| 38 \ \Omega = \frac{47 \ \Omega \cdot 38 \ \Omega}{47 \ \Omega + 38 \ \Omega} = 21.0118 \ \Omega$$

$$R_{EQ} = R_1 \| 21.0112 \ \Omega = 15 \ \Omega \| 21.0118 \ \Omega$$

$$R_{EQ} = \frac{15 \ \Omega \cdot 21.0112 \ \Omega}{15 \ \Omega + 21.0112 \ \Omega} = 8.7520 \ \Omega$$

8.

$$R_3 \| R_4 = 4.7 \text{ M} \| 2.2 \text{ M} = \frac{4.7 \text{ M} \cdot 2.2 \text{ M}}{4.7 \text{ M} + 2.2 \text{ M}} = 1.4986 \text{ M}$$

$$R_{EQ} = R_1 + R_2 \| 1.4986 \text{ M} = (0.950 \text{ M} + 1.2 \text{ M}) \| 1.4986 \text{ M}$$

$$R_{EQ} = \frac{2.15 \text{ M} \cdot 1.4986 \text{ M}}{2.15 \text{ M} + 1.4986 \text{ M}} = 0.8831 \text{ M}\Omega$$

Multi-Resistor Calculations 1

1.

$$R_{EQ} = 1 \text{ k} + 2 \text{ k} = 3 \text{ k}\Omega$$

$$I = \frac{V_B}{R_{EQ}} = \frac{6 \text{ V}}{3 \text{ k}} = 2.0000 \text{ mA}$$

$$V_1 = IR_1 = 2.0000 \text{ mA} \cdot 1 \text{ k} = 2.0000 \text{ V}$$

2.

$$I = \frac{V_B}{R_2} = \frac{9 \text{ V}}{8 \text{ k}\Omega} = 1.1250 \text{ mA}$$

3.

$$R_2 \| R_3 = 10 \text{ k} \| 5 \text{ k} = \frac{10 \text{ k} \cdot 5 \text{ k}}{10 \text{ k} + 5 \text{ k}} = 3.3333 \text{ k}$$

$$R_{EQ} = R_1 + 3.3333 \text{ k} = 5 \text{ k} + 3.3333 \text{ k} = 8.3333 \text{ k}\Omega$$

$$I_1 = \frac{V_B}{R_{EQ}} = \frac{9 \text{ V}}{8.3333 \text{ k}} = 1.0800 \text{ mA}$$

$$V_1 = I_1 R_1 = 1.0800 \text{ mA} \cdot 5 \text{ k} = 5.4000 \text{ V}$$

$$V_3 = 9 \text{ V} - V_1 = 9 \text{ V} - 5.4000 \text{ V} = 3.6000 \text{ V}$$

$$I_3 \text{ (through } R_3) = \frac{V_3}{R_3} = \frac{3.6000 \text{ V}}{5 \text{ k}} = 0.7200 \text{ mA}$$

$$P_{R3} = V_3 I_3 = 2.5920 \text{ mW}$$

4.

$$R_2 \| R_3 = \frac{2 \text{ k} \cdot 2 \text{ k}}{2 \text{ k} + 2 \text{ k}} = 1 \text{ k}$$

$$R_{EQ} = R_1 + 1 \text{ k} + R_4 = 2 \text{ k} + 1 \text{ k} + 4 \text{ k} = 7 \text{ k}\Omega$$

$$I_1 = \frac{V_B}{R_{EQ}} = \frac{6 \text{ V}}{7 \text{ k}} = 0.8571 \text{ mA}$$

$$V_4 = I_1 R_4 = 0.8571 \text{ mA} \cdot 4 \text{ k} = 3.4284 \text{ V}$$

Multi-Resistor Calculations 2

1.

$$R_2 \| R_3 = \frac{2.2\text{ k} \cdot 4.5\text{ k}}{2.2\text{ k} + 4.5\text{ k}} = 1.4776\text{ k}$$

$$R_{EQ} = R_1 + 1.4776\text{ k} = 1.5\text{ k} + 1.4776\text{ k} = 2.9776\text{ k}\Omega$$

$$I_1 = \frac{V_B}{R_{EQ}} = \frac{5\text{ V}}{2.9776\text{ k}} = 1.6792\text{ mA}$$

$$V_1 = I_1 R_1 = 1.6792\text{ mA} \cdot 1.5\text{ k} = 2.5188\text{ V}$$

$$V_2 = V_B - V_1 = 5\text{ V} - 2.5188\text{ V} = 2.4812\text{ V}$$

$$I_{R2} = \frac{V_2}{R_2} = \frac{2.4812\text{ V}}{2.2\text{ k}} = 1.1278\text{ mA}$$

2.

$$R_2 + R_3 + R_4 = 0.9\text{ k} + 1\text{ k} + 2.1\text{ k} = 4\text{ k}$$

$$R_{EQ} = 4.3\text{ k} \| 4\text{ k} = \frac{4.3\text{ k} \cdot 4\text{ k}}{4.3\text{ k} + 4\text{ k}} = 2.0723\text{ k}\Omega$$

$$I_1 = \frac{V_B}{R_{EQ}} = \frac{9\text{ V}}{2.0723\text{ k}\Omega} = 4.3430\text{ mA}$$

$$I_{R1} = \frac{V_B}{R_1} = \frac{9\text{ V}}{4.3\text{ k}} = 2.0930\text{ mA}$$

$$I_{R2} = I_1 - I_{R1} = 4.3430\text{ mA} - 2.0930\text{ mA} = 2.2500\text{ mA}$$

$$V_3 = I_{R2} R_3 = 2.2500\text{ mA} \cdot 1.0\text{ k} = 2.2500\text{ V}$$

3.

$$R_3 + R_4 = 3.3\text{ k} + 4.7\text{ k} = 8\text{ k}$$

$$R_2 \| 8\text{ k} = \frac{2\text{ k} \cdot 8\text{ k}}{2\text{ k} + 8\text{ k}} = 1.6\text{ k}$$

$$R_{EQ} = R_1 + 1.6\text{ k} = 0.5\text{ k} + 1.6\text{ k} = 2.1000\text{ k}$$

$$I_1 = \frac{V_B}{R_{EQ}} = \frac{12\text{ V}}{2.1000\text{ k}} = 5.7143\text{ mA}$$

$$V_1 = I_1 R_1 = 5.7143\text{ mA} \cdot 0.5\text{ k} = 2.8572\text{ V}$$

$$V_2 = V_B - V_1 = 12\text{ V} - 2.8572\text{ V} = 9.1428\text{ V}$$

$$I_{R2} = \frac{V_2}{R_2} = \frac{9.1428\text{ V}}{2\text{ k}} = 4.5714\text{ mA}$$

$$I_{R3} = I_1 - I_{R2} = 5.7143\text{ mA} - 4.5714\text{ mA} = 1.1429\text{ mA}$$

$$V_3 = I_{R3} R_3 = 1.1429\text{ mA} \cdot 3.3\text{ k} = 3.772\text{ V}$$

$$V_4 = V_B - V_1 - V_3 = 12\text{ V} - 2.8572\text{ V} - 3.772\text{ V} = 5.3708\text{ V}$$

$$P_{R4} = V_4 I_{R3} = 5.3708\text{ V} \cdot 1.1429\text{ mA} = 6.1383\text{ mW}$$

4.

$R_3 + R_4 + R_5 = 1.5 \text{ M} + 1.5 \text{ M} + 3.3 \text{ M} = 6.3 \text{ M}$

$R_2 \| 6.3 \text{ M} = \dfrac{4.7 \text{ M} \cdot 6.3 \text{ M}}{4.7 \text{ M} + 6.3 \text{ M}} = 2.6918 \text{ M}$

$R_{EQ} = R_1 + 2.6918 \text{ M} + R_8 = 1.5 \text{ M} + 2.6918 \text{ M} + 3.3 \text{ M} = 7.4918 \text{ M}\Omega$

$V_B = I_1 R_{EQ}$

$I_1 = \dfrac{V_B}{R_{EQ}} = \dfrac{6 \text{ V}}{7.4918 \text{ M}\Omega} = 0.8009 \text{ μA}$

$V_1 = I_1 R_1 = 0.8009 \text{ μA} \cdot 1.5 \text{ M} = 1.2013 \text{ V}$

$V_6 = I_1 R_6 = 0.8009 \text{ μA} \cdot 3.3 \text{ M} = 2.6430 \text{ V}$

$V_B = V_1 + V_2 + V_6$

$V_2 = V_B - V_1 - V_6 = 6 \text{ V} - 1.2013 \text{ V} - 2.6430 \text{ V} = 2.1557 \text{ V}$

$V_2 = I_2 R_2$

$I_{R2} = \dfrac{V_2}{R_2} = \dfrac{2.1557 \text{ V}}{4.7 \text{ M}} = 0.4587 \text{ μA}$

$I_{R1} = I_{R2} + I_{R3}$

$I_{R3} = I_1 - I_{R2} = 0.8009 \text{ μA} - 0.4587 \text{ μA} = 0.3422 \text{ μA}$

$V_3 = I_{R3} R_3 = 0.3422 \text{ μA} \cdot 1.5 \text{ M} = 0.5133 \text{ V}$

$V_4 = I_{R3} R_4 = 0.3422 \text{ μA} \cdot 1.5 \text{ M} = 0.5133 \text{ V}$

$V_5 = I_{R3} R_5 = 0.3422 \text{ μA} \cdot 3.3 \text{ M} = 1.1293 \text{ V}$

Multi-Resistor Calculations 3

1.

$R_3 \| R_4 = 3 \text{ k} \| 3 \text{ k} = 1.5 \text{ k}$

$R_{EQ} = R_1 + R_2 + 1.5 \text{ k} = 2 \text{ k} + 2 \text{ k} + 1.5 \text{ k} = 5.5 \text{ k}\Omega$

$I_1 = \dfrac{V_B}{R_{EQ}} = \dfrac{5.5 \text{ V}}{5.5 \text{ k}\Omega} = 1.0000 \text{ mA}$

$V_1 = I_1 R_1 = 1.0000 \text{ mA} \cdot 2 \text{ k} = 2.0000 \text{ V}$

$V_2 = I_1 R_2 = 1.0000 \text{ mA} \cdot 2 \text{ k} = 2.0000 \text{ V}$

$V_3 = V_B - V_1 - V_2 = 5.5 \text{ V} - 2.0000 \text{ V} - 2.0000 \text{ V} = 1.5000 \text{ V}$

$I_{R3} = \dfrac{V_3}{R_3} = \dfrac{1.5 \text{ V}}{3.0 \text{ k}\Omega} = 0.5000 \text{ mA}$

2.

$$R_{EQ} = (R_1 \| R_2) + R_3 = \frac{4.4 \text{ k} \cdot 5.3 \text{ k}}{4.4 \text{ k} + 5.3 \text{ k}} + 6.1 \text{ k} = 8.5041 \text{ k}\Omega$$

$$I_1 = \frac{V_B}{R_{EQ}} = \frac{8 \text{ V}}{8.5041 \text{ k}\Omega} = 0.9407 \text{ mA}$$

$$I_1 = I_{R3}$$

$$V_3 = I_{R3}R_3 = 0.9407 \text{ mA} \cdot 6.1 \text{ k}\Omega = 5.7383 \text{ V}$$

$$V_2 = V_B - V_3 = 8 \text{ V} - 5.7383 \text{ V} = 2.2617 \text{ V}$$

$$I_{R2} = \frac{V_2}{R_2} = \frac{2.2617 \text{ V}}{5.3 \text{ k}} = 0.4267 \text{ mA}$$

3.

$$(R_2 + R_3) \| R_5 = 0.8 \text{ k} \| 0.51 \text{ k} = \frac{0.8 \text{ k} \cdot 0.51 \text{ k}}{0.8 \text{ k} + 0.51 \text{ k}} = 0.3115 \text{ k}$$

$$R_{EQ} = R_1 + 0.3114 \text{ k} + R_4 = 0.77 \text{ k} + 0.3115 \text{ k} + 0.87 \text{ k} = 1.9515 \text{ k}\Omega$$

$$I_1 = \frac{V_B}{R_{EQ}} = \frac{12 \text{ V}}{1.9514 \text{ k}\Omega} = 6.1491 \text{ mA}$$

$$V_1 = I_1 R_1 = 6.1491 \text{ mA} \cdot 0.77 \text{ k} = 4.7348 \text{ V}$$

$$V_4 = I_1 R_4 = 6.1491 \text{ mA} \cdot 0.87 \text{ k} = 5.3497 \text{ V}$$

$$V_5 = V_B - V_1 - V_4 = 12 \text{ V} - 4.7348 \text{ V} - 5.3497 \text{ V} = 1.9155 \text{ V}$$

$$I_3 \left(\text{Through } R_5\right) = \frac{V_5}{R_5} = \frac{1.9155 \text{ V}}{0.51 \text{ k}\Omega} = 3.7559 \text{ mA}$$

$$I_2 = I_1 - I_3 = 6.1491 \text{ mA} - 3.7559 \text{ mA} = 2.3932 \text{ mA}$$

$$V_2 = I_{R2}R_2 = 2.3932 \text{ mA} \cdot 0.33 \text{ k} = 0.7898 \text{ V}$$

$$P_{R2} = V_2 I_{R2} = 0.7898 \text{ V} \cdot 2.3932 \text{ mA} = 1.8901 \text{ mW}$$

$$V_3 = I_{R3}R_3 = 2.3932 \text{ mA} \cdot 0.47 \text{ k} = 1.1248 \text{ V}$$

$$P_{R3} = V_3 I_{R3} = 1.1248 \text{ V} \cdot 2.3932 \text{ mA} = 2.6919 \text{ mW}$$

4.

$R_3 + R_4 + R_5 = 2.4\text{ k} + 5.1\text{ k} + 4.7\text{ k} = 12.2\text{ k}$

$R_2 \parallel 12.2\text{ k} = \dfrac{1.3\text{ k} \cdot 12.2\text{ k}}{1.3\text{ k} + 12.2\text{ k}} = 1.1748\text{ k}$

$R_{EQ} = R_1 \parallel 1.1748\text{ k} = \dfrac{3.3\text{ k} \cdot 1.1748\text{ k}}{3.3\text{ k} + 1.1748\text{ k}} = 0.8664\text{ k}\Omega$

$I_1 = \dfrac{V_B}{R_{EQ}} = \dfrac{4.7\text{ V}}{0.8664\text{ k}\Omega} = 5.4247\text{ mA}$

$I_{R1} = \dfrac{V_B}{R_1} = \dfrac{4.7\text{ V}}{3.3\text{ k}\Omega} = 1.4242\text{ mA}$

$I_{R2} = \dfrac{V_B}{R_2} = \dfrac{4.7\text{ V}}{1.3\text{ k}\Omega} = 3.6154\text{ mA}$

$I_{R5} = I_1 - I_{R1} - I_{R2} = 5.4247\text{ mA} - 1.4242\text{ mA} - 3.6154\text{ mA} = 0.3851\text{ mA}$

$V_5 = I_{R5}R_5 = 0.3851\text{ mA} \cdot 4.7\text{ k} = 1.8100\text{ V}$

$P_{R5} = V_5 I_{R5} = 1.8100\text{ V} \cdot 0.3851\text{ mA} = 0.6970\text{ mW}$

Chapter 10

Solubility Calculations

1.

$$0.32 \ \frac{\text{g}}{\text{mL}} = \frac{3500 \text{ g}}{(x) \text{ mL}}$$

$$(x) \text{ mL} \cdot 0.32 \ \frac{\text{g}}{\text{mL}} = 3500 \text{ g}$$

$$(x) \text{ mL} = \frac{3500 \text{ g}}{0.32 \ \frac{\text{g}}{\text{mL}}} = 11{,}000 \text{ mL} \cdot \frac{1 \text{ L}}{1000 \text{ mL}} = 11 \text{ L}$$

2.

$$1.01 \ \frac{\text{g}}{\text{mL}} \cdot \frac{1 \text{ kg}}{1000 \text{ g}} \cdot \frac{1000 \text{ mL}}{1 \text{ L}} \cdot \frac{1000 \text{ L}}{1 \text{ m}^3} = 1010 \ \frac{\text{kg}}{\text{m}^3}$$

$$1010 \ \frac{\text{kg}}{\text{m}^3} = \frac{(x) \text{ kg}}{5.00 \text{ m}^3}$$

$$1010 \text{ kg} \cdot 5.00 = (x) \text{ kg} = 5050 \text{ kg}$$

3.

$$\frac{1.750 \text{ kg}}{3.66 \text{ L}} \cdot \frac{1 \text{ L}}{1000 \text{ mL}} \cdot \frac{1000 \text{ g}}{1 \text{ kg}} = 0.478 \ \frac{\text{g}}{\text{mL}}$$

4.

$$\frac{3610 \text{ kg}}{45{,}550 \text{ L}} \cdot \frac{1000 \text{ L}}{1 \text{ m}^3} = 79.3 \ \frac{\text{kg}}{\text{m}^3}$$

5.

$$79.3 \ \frac{\text{kg}}{\text{m}^3} = \frac{4.68 \times 10^7 \text{ kg}}{(x) \text{ m}^3}$$

$$(x) \ 79.3 \text{ kg} = 1 \text{ m}^3 \cdot 4.68 \times 10^7 \text{ kg}$$

$$(x) = \frac{4.68 \times 10^7 \text{ kg}}{79.3 \text{ kg}} = 590{,}164 \text{ m}^3 \cdot \frac{1000 \text{ L}}{1 \text{ m}^3} \cdot \frac{1 \text{ ML}}{10^6 \text{ L}} = 5.90 \times 10^2 \text{ ML}$$

Chapter 11

Volume, Mass, and Weight Exercises

1.

$$98.34 \ \frac{kg}{m^3} \cdot \frac{1000 \ g}{1 \ kg} \cdot \left(\frac{1 \ m}{100 \ cm}\right)^3 = 0.09834 \ \frac{g}{cm^3}$$

2.

$$42 \ mL \cdot \frac{1 \ L}{1000 \ mL} \cdot \frac{1 \ gal}{3.785 \ L} = 0.011 \ gal$$

3.

$$F_w = 18.5 \ lb \cdot \frac{4.45 \ N}{1 \ lb} = 82.3 \ N$$

$$m = ?$$

$$F_w = mg$$

$$m = \frac{F_w}{g} = \frac{82.3 \ N}{9.80 \ \dfrac{m}{s^2}} = 8.40 \ kg$$

4.

$$3.6711 \times 10^4 \ \frac{g}{mL} \cdot \frac{1 \ kg}{1000 \ g} \cdot \frac{1000 \ mL}{1 \ L} \cdot \frac{1000 \ L}{1 \ m^3} = 3.6711 \times 10^7 \ \frac{kg}{m^3}$$

5.

$$1.957 \times 10^4 \ in^3 \cdot \left(\frac{2.54 \ cm}{1 \ in}\right)^3 = 320{,}700 \ cm^3$$

6.

$$455 \ mL \cdot \frac{1 \ L}{1000 \ mL} \cdot \frac{1 \ m^3}{1000 \ L} = 0.000455 \ m^3$$

7.

$m = 46,000$ kg

$F_w = ?$

$F_w = mg = 46,000$ kg $\cdot 9.80 \ \dfrac{m}{s^2} = 450,000$ N

4.508×10^5 N $\cdot \dfrac{1 \text{ lb}}{4.45 \text{ N}} = 1.0 \times 10^5$ lb

---

8.

32.11 L $\cdot \dfrac{1000 \text{ cm}^3}{1 \text{ L}} \cdot \left(\dfrac{1 \text{ in}}{2.54 \text{ cm}} \right)^3 = 1959$ in^3

9.

$F_w = 14.89$ N $\cdot \dfrac{1 \text{ lb}}{4.45 \text{ N}} = 3.35$ lb

$m = ?$

$m = \dfrac{F_w}{g} = \dfrac{14.89 \text{ N}}{9.80 \ \dfrac{m}{s^2}} = 1.52$ kg

---

10.

36.00 cm$^3 \cdot \left(\dfrac{1 \text{ m}}{100 \text{ cm}} \right)^3 = 3.6 \times 10^{-5}$ m^3

11.

9.11 m$^3 \cdot \left(\dfrac{100 \text{ cm}}{1 \text{ m}} \right)^3 = 9.11 \times 10^6$ cm^3

12.

4.11×10^5 m$^3 \cdot \dfrac{1000 \text{ L}}{1 \text{ m}^3} = 4.11 \times 10^8$ L

13.

$F_w = 55,789$ lb $\cdot \dfrac{4.45 \text{ N}}{1 \text{ lb}} = 2.48261 \times 10^5$ N

$m = \dfrac{F_w}{g} = \dfrac{2.48261 \times 10^5 \text{ N}}{9.80 \ \dfrac{m}{s^2}} = 25,300$ kg

---

14.

$$5.022 \ \frac{g}{cm^3} \cdot \frac{1 \text{ kg}}{1000 \text{ g}} \cdot \left(\frac{100 \text{ cm}}{1 \text{ m}} \right)^3 = 5022 \ \frac{kg}{m^3}$$

15.

$$3.76 \times 10^{-4} \ \frac{g}{mL} \cdot \frac{1000 \text{ mL}}{1 \text{ L}} \cdot \frac{1 \text{ L}}{1000 \text{ cm}^3} = 3.76 \times 10^{-4} \ \frac{g}{cm^3}$$

16.

$$F_w = 50,000 \text{ N} \cdot \frac{1 \text{ lb}}{4.45 \text{ N}} = 10,000 \text{ lb}$$

$$m = \frac{F_w}{g} = \frac{50,000 \text{ N}}{9.80 \ \frac{m}{s^2}} = 5000 \text{ kg}$$

17.

$$1.75 \times 10^{-6} \text{ m}^3 \cdot \left(\frac{100 \text{ cm}}{1 \text{ m}} \right)^3 = 1.75 \text{ cm}^3$$

18.

$$100.5 \text{ ft}^3 \cdot \left(\frac{12 \text{ in}}{1 \text{ ft}} \right)^3 \cdot \left(\frac{2.54 \text{ cm}}{1 \text{ in}} \right)^3 \cdot \left(\frac{1 \text{ m}}{100 \text{ cm}} \right)^3 = 2.846 \text{ m}^3$$

19.

$$37 \text{ m}^3 \cdot \left(\frac{100 \text{ cm}}{1 \text{ m}} \right)^3 \cdot \left(\frac{1 \text{ in}}{2.54 \text{ cm}} \right)^3 = 2,300,000 \text{ in}^3$$

20.

$$750 \text{ cm}^3 \cdot \frac{1 \text{ L}}{1000 \text{ cm}^3} = 0.75 \text{ L}$$

21.

$$5,755,000 \text{ gal} \cdot \frac{3.785 \text{ L}}{1 \text{ gal}} \cdot \frac{1000 \text{ cm}^3}{1 \text{ L}} \cdot \left(\frac{1 \text{ m}}{100 \text{ cm}} \right)^3 = 21,780 \text{ m}^3$$

Density Exercises

1.

$m = 0.196$ g

$V = 100.1$ mL

$\rho = ?$

$$\rho = \frac{m}{V} = \frac{0.196 \text{ g}}{100.1 \text{ mL}} = 1.96 \times 10^{-3} \frac{\text{g}}{\text{mL}}$$

2.

$$\rho = 955 \frac{\text{kg}}{\text{m}^3} \cdot \frac{1000 \text{ g}}{1 \text{ kg}} \cdot \left(\frac{1 \text{ m}}{100 \text{ cm}}\right)^3 \cdot \frac{100 \text{ cm}^3}{1 \text{ L}} \cdot \frac{1 \text{ L}}{1000 \text{ mL}} = 0.955 \frac{\text{g}}{\text{mL}}$$

$m = 550$ g

$V = ?$

$$\rho = \frac{m}{V}$$

$$V = \frac{m}{\rho} = \frac{550 \text{ g}}{0.955 \frac{\text{g}}{\text{mL}}} = 580 \text{ mL}$$

3.

$$m = 15.7 \text{ kg} \cdot \frac{1000 \text{ g}}{1 \text{ kg}} = 1.57 \times 10^4 \text{ g}$$

$$\rho = 5.32 \frac{\text{g}}{\text{cm}^3}$$

$V = ?$

$$\rho = \frac{m}{V}$$

$$V = \frac{m}{\rho} = \frac{1.57 \times 10^4 \text{ g}}{5.32 \frac{\text{g}}{\text{cm}^3}} = 2,950 \text{ cm}^3$$

$$2.95 \times 10^3 \text{ cm}^3 \cdot \left(\frac{1 \text{ m}}{100 \text{ cm}}\right)^3 = 0.00295 \text{ m}^3$$

4.

$l = 3.00$ cm

$w = 3.00$ cm

$h = 3.00$ cm

$F_w = 5.336 \times 10^{-2} \text{ lb} \cdot \dfrac{4.45 \text{ N}}{1 \text{ lb}} = 2.375 \times 10^{-1} \text{ N}$

$\rho = ?$

$F_w = mg$

$m = \dfrac{F_w}{g} = \dfrac{2.375 \times 10^{-1} \text{ N}}{9.80 \ \dfrac{\text{m}}{\text{s}^2}} = 2.423 \times 10^{-2} \text{ kg} \cdot \dfrac{1000 \text{ g}}{1 \text{ kg}} = 24.23 \text{ g}$

$V = l \cdot w \cdot h = (3.00 \text{ cm})^3 = 27 \text{ cm}^3$

$\rho = \dfrac{m}{V} = \dfrac{24.23 \text{ g}}{27 \text{ cm}^3} = 0.897 \ \dfrac{\text{g}}{\text{cm}^3}$

5.

$V_1 = 23.35$ mL

$V_2 = 27.79$ mL

$m = 32.1$ g

$\rho = ?$

$V_{Total} = 27.79 \text{ mL} - 23.35 \text{ mL} = 4.44 \text{ mL} \cdot \dfrac{1 \text{ mL}}{1 \text{ cm}^3} = 4.44 \text{ cm}^3$

$\rho = \dfrac{32.1 \text{ g}}{4.44 \text{ cm}^3} = 7.23 \ \dfrac{\text{g}}{\text{cm}^3}$

6.

$$h = 34.5 \text{ in} \cdot \frac{1 \text{ ft}}{12 \text{ in}} \cdot \frac{0.3048 \text{ m}}{1 \text{ ft}} = 0.8763 \text{ m}$$

$$d = 24 \text{ in} \cdot \frac{1 \text{ ft}}{12 \text{ in}} \cdot \frac{0.3048 \text{ m}}{1 \text{ ft}} = 0.6096 \text{ m}$$

$$\rho = 810 \frac{\text{kg}}{\text{m}^3}$$

$$m = ?$$

$$V = \pi r^2 h = \pi \cdot \left(\frac{0.6096 \text{ m}}{2} \right)^2 \cdot 0.2558 \text{ m}^3$$

$$\rho = \frac{m}{V}$$

$$m = V\rho = 0.2558 \text{ m}^3 \cdot 810 \frac{\text{kg}}{\text{m}^3} = 210 \text{ kg}$$

8.

Ethanol

$$V = 5 \text{ mL}$$

$$m = 3.9 \text{ g}$$

$$\rho = \frac{m}{V} = \frac{3.9 \text{ g}}{5 \text{ mL}} = 0.78 \frac{\text{g}}{\text{mL}}$$

Benzene

$$V = 7.5 \text{ L} \cdot \frac{1000 \text{ mL}}{\text{L}} = 7500 \text{ mL}$$

$$m = 6.6 \text{ kg} \cdot \frac{1000 \text{ g}}{\text{kg}} = 6600 \text{ g}$$

$$\rho = \frac{m}{V} = \frac{6600 \text{ g}}{7500 \text{ mL}} = 0.88 \frac{\text{g}}{\text{mL}}$$

9.

$$\rho = 7{,}830 \ \frac{\text{kg}}{\text{m}^3} \cdot \frac{1000 \ \text{g}}{1 \ \text{kg}} \cdot \left(\frac{1 \ \text{m}}{100 \ \text{cm}}\right)^3 = 7.83 \ \frac{\text{g}}{\text{cm}^3}$$

$l = 2.1 \ \text{cm}$

$w = 3.5 \ \text{cm}$

$m = 94.5 \ \text{g}$

$h = ?$

$$\rho = \frac{m}{V}$$

$$V = \frac{m}{\rho} = \frac{94.5 \ \text{g}}{7.83 \ \dfrac{\text{g}}{\text{cm}^3}} = 12.07 \ \text{cm}^3$$

$$V = l \cdot w \cdot h$$

$$h = \frac{V}{l \cdot w} = \frac{12.07 \ \text{cm}^3}{2.1 \ \text{cm} \cdot 3.5 \ \text{cm}} = 1.6 \ \text{cm}$$

10.

$$m = 306 \ \text{g} \cdot \frac{1 \ \text{kg}}{1000 \ \text{g}} = 0.306 \ \text{kg}$$

$$V = 22.5 \ \text{mL} \cdot \frac{1 \ \text{cm}^3}{1 \ \text{mL}} \cdot \left(\frac{1 \ \text{m}}{100 \ \text{cm}}\right)^3 = 2.25 \times 10^{-5} \ \text{m}^3$$

$\rho = ?$

$$\rho = \frac{m}{V} = \frac{0.306 \ \text{kg}}{2.25 \times 10^{-5} \ \text{m}^3} = 13{,}600 \ \frac{\text{kg}}{\text{m}^3}$$

11.

$$\rho = 10.501 \ \frac{g}{cm^3}$$

$$h = 4.500 \ cm$$

$$d = 2.7500 \ cm$$

$$F_w = ?$$

$$V = \pi r^2 h = \pi \cdot \left(\frac{2.7500 \ cm}{2} \right)^2 \cdot 4.500 \ cm = 26.728 \ cm^3$$

$$\rho = \frac{m}{V}$$

$$m = V\rho = 26.728 \ cm^3 \cdot 10.501 \ \frac{g}{cm^3} = 280.67 \ g \cdot \frac{1 \ kg}{1000 \ g} = 0.28067$$

$$F_w = mg = 0.28067 \cdot 9.80 \ \frac{m}{s^2} = 2.7506 \ N \cdot \frac{1 \ lb}{4.45 \ N} = 0.618 \ lb$$

12.

$$\rho = 1{,}000.0 \ \frac{kg}{m^3}$$

$$V = 5.6 \ L \cdot \frac{1000 \ cm^3}{1 \ L} \cdot \left(\frac{1 \ m}{100 \ cm} \right)^3 = 5.6 \times 10^{-3} \ m^3$$

$$m = ?$$

$$\rho = \frac{m}{V}$$

$$m = V\rho = 5.6 \times 10^{-3} \ m^3 \cdot 1{,}000.0 \ \frac{kg}{m^3} = 5.6 \ kg$$

13.

$$V = 3.0 \times 10^6 \text{ gal} \cdot \frac{3.785 \text{ L}}{1 \text{ gal}} \cdot \frac{1 \text{ m}^3}{1000 \text{ L}} = 1.14 \times 10^4 \text{ m}^3$$

$$\rho = 998 \frac{\text{kg}}{\text{m}^3}$$

$$F_w = ?$$

$$\rho = \frac{m}{V}$$

$$m = V\rho = 1.14 \times 10^4 \text{ m}^3 \cdot 998 \frac{\text{kg}}{\text{m}^3} = 1.14 \times 10^7 \text{ kg}$$

$$F_w = mg = 1.14 \times 10^7 \text{ kg} \cdot 9.80 \frac{\text{m}}{\text{s}^2} = 1.12 \times 10^8 \text{ N}$$

$$F_w = 1.12 \times 10^8 \text{ N} \cdot \frac{1 \text{ lb}}{4.45 \text{ N}} = 25,000,000 \text{ lb}$$

15.

$$\rho = 7.85 \frac{\text{g}}{\text{cm}^3} \cdot \frac{1 \text{ kg}}{1000 \text{ g}} \cdot \left(\frac{100 \text{ cm}}{1 \text{ m}} \right)^3 = 7,850 \frac{\text{kg}}{\text{m}^3}$$

$$d = 6.1875 \text{ in} \cdot \frac{1 \text{ ft}}{12 \text{ in}} \cdot \frac{0.3048 \text{ m}}{1 \text{ ft}} = 0.15716 \text{ m}$$

$$h = 7.0000 \text{ in} \cdot \frac{1 \text{ ft}}{12 \text{ in}} \cdot \frac{0.3048 \text{ m}}{1 \text{ ft}} = 0.17780 \text{ m}$$

$$F_w = ?$$

$$V = \pi r^2 h = \pi \cdot \left(\frac{0.15716 \text{ m}}{2} \right)^2 \cdot 0.1778 \text{ m} = 0.0034491 \text{ m}^3$$

$$\rho = \frac{m}{V}$$

$$m = V\rho = 0.0034491 \text{ m}^3 \cdot 7,850 \frac{\text{kg}}{\text{m}^3} = 27.075 \text{ kg}$$

$$F_w = mg = 27.075 \text{ kg} \cdot 9.80 \frac{\text{m}}{\text{s}^2} = 265.34 \text{ N} \cdot \frac{1 \text{ lb}}{4.45 \text{ N}} = 59.6 \text{ lb}$$

16.

$$\rho = 160 \ \frac{kg}{m^3}$$

$$d = 2.0 \ ft \cdot \frac{0.3048 \ m}{1 \ ft} = 0.6096 \ m$$

$$h = 45 \ ft \cdot \frac{0.3048 \ m}{1 \ ft} = 13.716 \ m$$

$$V = \pi r^2 h = \pi \cdot \left(\frac{0.6096 \ m}{2} \right)^2 \cdot 13.716 \ m = 4.0032 \ m^3$$

$$\rho = \frac{m}{V}$$

$$m = V\rho = 4.0032 \ m^3 \cdot 160 \ \frac{kg}{m^3} = 640.51 \ kg$$

$$F_w = mg = 640.51 \ kg \cdot 9.80 \ \frac{m}{s^2} = 6{,}277.0 \ N$$

$$9 \ trees \cdot 6{,}277.0 \ \frac{N}{tree} \cdot \frac{1 \ lb}{4.45 \ N} = 13{,}000 \ lb$$

17.

$$l = 50.0 \ m$$

$$w = 25.0 \ m$$

$$d = 2.00 \ m$$

$$\rho = 998 \ \frac{kg}{m^3}$$

$$V = ?$$

$$F_w = ?$$

$$V = lwd = 50.0 \ m \cdot 25.0 \ m \cdot 2.00 \ m = 2500.0 \ m^3 \cdot \frac{1000 \ L}{1 \ m^3} \cdot \frac{1 \ gal}{3.785 \ L} = 6.61 \times 10^5 \ gal$$

$$\rho = \frac{m}{V}$$

$$m = V\rho = 2500.0 \ m^3 \cdot 998 \ \frac{kg}{m^3} = 2.495 \times 10^6 \ kg$$

$$F_w = mg = 2.495 \times 10^6 \ kg \cdot 9.80 \ \frac{m}{s^2} = 2.4451 \times 10^7 \ N \cdot \frac{1 \ lb}{4.45 \ N} \cdot \frac{1 \ ton}{2000 \ lb} = 2750 \ tons$$

Chapter 13

3.

$5 \cdot 1.673 \times 10^{-27} \text{ kg} = 8.365 \times 10^{-27} \text{ kg}$

$5 \cdot 9.11 \times 10^{-31} \text{ kg} = 4.555 \times 10^{-30} \text{ kg}$

$6 \cdot 1.675 \times 10^{-27} \text{ kg} = 1.005 \times 10^{-26} \text{ kg}$

$8.365 \times 10^{-27} \text{ kg} + 4.555 \times 10^{-30} \text{ kg} + 1.005 \times 10^{-26} \text{ kg} = 1.8419555 \times 10^{-26} \text{ kg}$